# Trees and Weather

## FOSS
**FULL OPTION SCIENCE SYSTEM**

**3rd Edition**

Developed at the Lawrence Hall of Science, University of California, Berkeley
Published and Distributed by Delta Education

**FOSS Lawrence Hall of Science Team**

Larry Malone and Linda De Lucchi, FOSS Project Codirectors and Lead Developers

Kathy Long, FOSS Assessment Director; David Lippman, Project Specialist and Editor; Carol Sevilla, Publications Design Coordinator; Rose Craig, Illustrator; Susan Stanley, Graphic Production; John Quick, Photographer; Trudihope Schlomowitz, Barbara Clinton, Editors; DeSondra Ward, Office Manager

**FOSS Curriculum Developers:** Brian Campbell, Teri Dannenberg, Alan Gould, Susan Kaschner Jagoda, Ann Moriarty, Jessica Penchos, Kimi Hosoume, Virginia Reid, Terry Shaw, Joanna Snyder, Erica Beck Spencer, Diana Velez, Natalie Yakushiji

**FOSS Technology Developers:** Susan Ketchner, Habiba Noor, Arzu Orgad

**FOSS Multimedia Team:** Kate Jordan, Multimedia Director; Nicole Medina, Senior Multimedia Producer; Matthew Jacoby, Lead Programmer; Geoffrey Thomas, Multimedia Programmer and Designer; Chris Linden, Designer; Chris Hamamoto, Designer; Dan Bluestein, Programmer; Roger Vang, Programmer; Christopher Cianciarulo, Programmer

**Delta Education Team**

Bonnie A. Piotrowski, FOSS Editorial Director

**Project Team:** Mathew Bacon, Jennifer Apt, Sandra Burke, Diane Gonciarz, Tom Guetling, Joann Hoy, Jacquelyn Lachance, Lisa Lachance, John Prescott

**Thank you to all FOSS Grades K-6 Trial Teachers**

Heather Ballard, Wilson Elementary, Coppell, TX; Mirith Ballestas De Barroso, Treasure Forest Elementary, Houston, TX; Terra L. Barton, Harry McKillop Elementary, Melissa, TX; Rhonda Bernard, Frances E. Norton Elementary, Allen, TX; Theresa Bissonnette, East Millbrook Magnet Middle School, Raleigh, NC; Heather Callaghan, Olive Chapel Elementary, Apex, NC; Katie Cannon, Las Colinas Elementary, Irving, TX; Kristy Cash, Wilson Elementary, Coppell, TX; Monica Coles, Swift Creek Elementary School, Raleigh, NC; Melissa Cook-Airhart, Harry McKillop Elementary, Melissa, TX; Hillary P. Croissant, Harry McKillop Elementary, Melissa, TX; Nancy Davis, Martha and Josh Morriss Mathematics and Engineering Elementary School, Texarkana, TX; Nancy Deveneau, Wilson Elementary, Coppell, TX; Karen Diaz, Las Colinas Elementary, Irving, TX; Marlana Dumas, Las Colinas Elementary, Irving, TX; Mary Evans, R.E. Good Elementary School, Carrollton, TX; Jacquelyn Farley, Moss Haven Elementary, Dallas, TX; Corinna Ferrier, Oak Forest Elementary, Humble, TX; Allison Fike, Wilson Elementary, Coppell, TX; Barbara Fugitt, Martha and Josh Morriss Mathematics and Engineering Elementary School, Texarkana, TX; Colleen Garvey, Farmington Woods Elementary, Cary, NC; Erin Gibson, Las Colinas Elementary, Irving, TX; Dollie Green, Melissa Ridge Intermediate School, Melissa, TX; Kim Hayes, Martha and Josh Morriss Mathematics and Engineering Elementary School, Texarkana, TX; Staci Lynn Hester, Lacy Elementary School, Raleigh, NC; Amanda Hill, Las Colinas Elementary, Irving, TX; Cindy Holder, Oak Forest Elementary, Humble, TX; Sarah Huber, Hodge Road Elementary, Knightdale, NC; Carol Kellum, Wallace Elementary, Dallas, TX; Brittani Kern, Fox Road Elementary, Raleigh, NC; Jodi Lay, Lufkin Road Middle School, Apex, NC; Ana Martinez, RISD Academy, Dallas, TX; Shaheen Mavani, Las Colinas Elementary, Irving, TX; Mary Linley McClendon, Math Science Technology Magnet School, Richardson, TX; Adam McKay, Davis Drive Elementary, Cary, NC; Anne Mechler, J. Erik Jonsson Community School, Dallas, TX; Shirley Diann Miller, The Rice School, Houston, TX; Anne Miller, J. Erik Jonsson Community School, Dallas, TX; Keri Minier, Las Colinas Elementary, Irving, TX; Stephanie Renee Nance, T.H. Rogers Elementary, Houston, TX; Elizabeth Noble, Las Colinas Elementary, Irving, TX; Courtney Noonan, Shadow Oaks Elementary School, Houston, TX; Sarah Peden, Aversboro Elementary School, Garner, NC; Carrie Prince, School at St. George Place, Houston, TX; Marlaina Pritchard, Melissa Ridge Intermediate School, Melissa, TX; Alice Pujol, J. Erik Jonsson Community School, Dallas, TX; Paul Rendon, Bentley Elementary, Oakland, CA; Janette Ridley, W.H. Wilson Elementary School, Coppell, TX; Kristina (Crickett) Roberts, W.H. Wilson Elementary School, Coppell, TX; Heather Rogers, Wendell Creative Arts & Science Magnet Elementary School, Wendell, NC; Megan Runion, Olive Chapel Elementary, Apex, NC; Christy Scheef, J. Erik Jonsson Community School, Dallas, TX; Samrawit Shawl, T.H. Rogers School, Houston, TX; Ashley Stephenson, J. Erik Jonsson Community School, Dallas, TX; Jolanta Stern, Browning Elementary School, Houston, TX; Gale Stimson, Bentley Elementary, Oakland, CA; Cathryn Sutton, Wilson Elementary, Coppell, TX; Brandi Swann, Westlawn Elementary School, Texarkana, TX; Robin Taylor, Arapaho Classical Magnet, Richardson, TX; Michael C. Thomas, Forest Lane Academy, Dallas, TX; Jomarga Thompkins, Lockhart Elementary, Houston, TX; Mary Timar, Madera Elementary, Lake Forest, CA; Helena Tongkeamha, White Rock Elementary, Dallas, TX; Linda Trampe, J. Erik Jonsson Community School, Dallas, TX; Charity VanHorn, Fred A. Olds Elementary, Raleigh, NC; Kathleen VanKeuren, Lufkin Road Middle School, Apex, NC; Megan Veron, Westwood Elementary School, Houston, TX; Mary Margaret Waters, Frances E. Norton Elementary, Allen, TX; Stephanie Robledo Watson, Ridgecrest Elementary School, Houston, TX; Lisa Webb, Madisonville Intermediate, Madisonville, TX; Nancy White, Canyon Creek Elementary, Austin, TX; Barbara Yurick, Oak Forest Elementary, Humble, TX; Linda Zittel, Mira Vista Elementary, Richmond, CA

**Photo Credits:** © bornholm/Shutterstock (cover); © Tatiana Grozetskaya/Shutterstock; © Laurie Meyer; © John Quick; © Erica Beck Spencer

**Published and Distributed by Delta Education, a member of the School Specialty Family**

The FOSS program was developed in part with the support of the National Science Foundation grant nos. MDR-8751727 and MDR-9150097. However, any opinions, findings, conclusions, statements, and recommendations expressed herein are those of the authors and do not necessarily reflect the views of NSF. FOSSmap was developed in collaboration between the BEAR Center at UC Berkeley and FOSS at the Lawrence Hall of Science. Score analysis is done through the BEAR Center Scoring Engine.

Trees and Weather — Teacher Toolkit, 1325300
Investigations Guide, 1391917
978-1-60902-668-4
Printing 2 – 6/2012
Webcrafters, Madison, WI

# Trees and Weather

## TABLE OF CONTENTS

Overview . . . . . . . . . . . . . . . . . . . . . . . . . . . . . . . . . 1

Materials . . . . . . . . . . . . . . . . . . . . . . . . . . . . . . . . 29

**Investigation 1: Observing Trees** . . . . . . . . . . . . . . . 41
Part 1: Observing Schoolyard Trees . . . . . . . . . . . . . . . 50
Part 2: Tree Parts . . . . . . . . . . . . . . . . . . . . . . . . . . . 62
Part 3: Tree Puzzles . . . . . . . . . . . . . . . . . . . . . . . . . 67
Part 4: Tree-Silhouette Cards . . . . . . . . . . . . . . . . . . . 71
Part 5: Adopt Schoolyard Trees . . . . . . . . . . . . . . . . . . 75
Part 6: A Tree Comes to Class . . . . . . . . . . . . . . . . . . 82

**Investigation 2: Observing Leaves** . . . . . . . . . . . . . . 93
Part 1: Leaf Walk . . . . . . . . . . . . . . . . . . . . . . . . . . 100
Part 2: Leaf Shapes . . . . . . . . . . . . . . . . . . . . . . . . . 106
Part 3: Comparing Leaves . . . . . . . . . . . . . . . . . . . . . 112
Part 4: Matching Leaf Silhouettes . . . . . . . . . . . . . . . . 117
Part 5: Leaf Books . . . . . . . . . . . . . . . . . . . . . . . . . . 122

**Investigation 3: Observing Weather** . . . . . . . . . . . . 131
Part 1: Weather Calendar . . . . . . . . . . . . . . . . . . . . . 138
Part 2: Recording Temperature . . . . . . . . . . . . . . . . . . 145
Part 3: Wind Direction . . . . . . . . . . . . . . . . . . . . . . . 151

**Investigation 4: Trees through the Seasons** . . . . . . 161
Part 1: Fall: What Comes from Trees? . . . . . . . . . . . . . 170
Part 2: Fall: Food from Trees . . . . . . . . . . . . . . . . . . . 173
Part 3: Fall: Visiting Adopted Trees . . . . . . . . . . . . . . . 177
Part 4: Winter: Evergreen Hunt . . . . . . . . . . . . . . . . . . 182
Part 5: Winter: Twigs . . . . . . . . . . . . . . . . . . . . . . . . 187
Part 6: Winter: Visiting Adopted Trees . . . . . . . . . . . . . 191
Part 7: Spring: Forcing Twigs . . . . . . . . . . . . . . . . . . . 195
Part 8: Spring: Bark Hunt . . . . . . . . . . . . . . . . . . . . . 198
Part 9: Spring: Visiting Adopted Trees . . . . . . . . . . . . . 202

## Contents

Introduction .............................1

Module Matrix .......................2

FOSS Conceptual Framework...4

Conceptual Framework in
Trees and Weather ..................6

FOSS Components ............... 10

FOSS Instructional Design ..... 12

FOSSweb and Technology ...... 20

Universal Design for
Learning................................ 22

Organizing the Classroom ...... 24

Safety in the Classroom
and Outdoors ........................ 26

Scheduling the Module .......... 27

FOSS K–8 Scope
and Sequence ....................... 28

## INTRODUCTION

The giant sequoia is the most massive living organism on Earth. It is a tree, magnificent in dimension and awe inspiring in its longevity and durability.

To a primary student, the oak on the corner, the pines at the park, and the mulberry tree at school are all giants. Systematic investigation of trees over the seasons will bring students to a better understanding of trees' place at school and in the community. Students will observe day-to-day changes in weather over the year, as well as the impact weather has on living things. The **Trees and Weather Module** provides students with solid experiences to help them know plants and their place on Earth. In this module, students will

- Observe and compare trees, using the senses.

- Observe and compare the shapes of leaves; compare leaf shapes to geometric shapes.

- Identify trees as resources that are used in everyday life.

- Observe weather by using senses and simple tools.

- Communicate observations made about different kinds of trees, leaves, and weather conditions orally and through drawings.

- Observe and record seasonal changes to living things.

| | Module Summary | Focus Questions |
|---|---|---|
| **Inv. 1: Observing Trees** | Students begin their study of trees by looking at the variety and structure of trees in the schoolyard. They work with representational materials to look more closely at the shapes of trees and their parts. They adopt schoolyard trees to observe changes through the year.  A living tree becomes part of the classroom for several weeks, and students complete the investigation by planting their class tree on the school grounds. | **What did we learn about our schoolyard trees?**<br>**What are the parts of trees?**<br>**What shapes are trees?**<br>**Which trees have similar shapes?**<br>**What can we find out about our adopted trees?**<br>**What  do trees need to grow?** |
| **Inv. 2: Observing Leaves** | Students begin with a schoolyard walk, focusing on the leaves of trees.  They match leaves with geometric shapes, go on a leaf hunt to compare properties of leaves, work at centers with representational materials, and make a leaf book. This investigation concludes with a story, *Our Very Own Tree*. | **What can we observe about leaves?**<br>**What shapes are leaves?**<br>**How are leaves different?**<br>**How are leaf edges different?** |
| **Inv. 3: Observing Weather** | Students share what they know about weather and how it relates to air.  A class weather monitor begins recording daily weather observations on a class calendar.  Students use weather pictures to indicate five basic types of weather.  They use a thermometer to measure relative temperature (how hot or cold it is) and make a wind sock to observe the wind direction and speed.  Students observe and compare objects in the sky during the day and at night. | **What is the weather today?**<br>**How can we measure the air temperature?**<br>**What does a wind sock tell us about the wind?** |
| **Inv. 4: Trees through the Seasons** | Students extend their understanding of trees as a growing, changing, living part of their world. During each season, students visit the schoolyard trees; observe their twigs, leaves, flowers, and seeds; and compare them to those from a previous season. | **What do fall trees look like?**<br>**What do winter trees look like?**<br>**What do spring trees look like?** |

| Content | Reading | Assessment |
|---|---|---|
| • Trees are living plants.<br>• Trees have structures: branches, leaves, trunk, and roots.<br>• Trees differ in size and shape.<br>• Trees have basic needs: light, air, nutrients, water, and space. | **Science Resources Book**<br>"Where Do Trees Grow?" | **Embedded Assessment**<br>Teacher observation |
| • Different kinds of trees have different leaves.<br>• Leaves have properties: size, shape, tip, edge, texture, and color.<br>• Leaves properties vary.<br>• Leaves can be described and compared by their properties. | **Books**<br>*How Do We Learn?*<br>*Our Very Own Tree* | **Embedded Assessment**<br>Teacher observation |
| • Weather is the condition in the air outdoors and can be described.<br>• Temperature is how hot or cold it is; thermometers measure temperature.<br>• Wind is moving air; a wind sock indicates wind direction and speed.<br>• Weather changes.<br>• The Sun, the Moon, and clouds are objects we see in the sky. | **Science Resources Book**<br>"Up in the Sky"<br>"Weather" | **Embedded Assessment**<br>Teacher observation |
| • Seasons change in a predictable annual pattern: fall, winter, spring, and summer.<br>• Bark, twigs, leaves, buds, flowers, fruits, and seeds are parts of trees.<br>• The buds on twigs grow into leaves or flowers.<br>• Trees change through the seasons.<br>• Some trees produce seeds that can grow into new trees of the same kind.<br>• Some trees lose their leaves in winter; others do not.<br>• Trees are living, growing plants. | **Science Resources Book**<br>"My Apple Tree"<br>"Orange Trees"<br>"Maple Trees" | **Embedded Assessment**<br>Teacher observation |

# FOSS CONCEPTUAL FRAMEWORK

FOSS has conceptual structure at the module level. The concepts are carefully selected and organized in a sequence that makes sense to students when presented as intended. In the last half decade, research has focused on learning progressions. The idea behind a learning progression is that **core ideas** in science are complex and wide-reaching—ideas such as the structure of matter or the relationship between the structure and function of organisms. From the age of awareness throughout life, matter and organisms are important to us. There are things we can and should understand about them in our primary school years, and progressively more complex and sophisticated things we should know about them as we gain experience and develop our cognitive abilities. When we can determine those logical progressions, we can develop meaningful and effective curriculum.

FOSS has elaborated learning progressions for core ideas in science for kindergarten through grade 6. Developing the learning progressions involves identifying successively more sophisticated ways of thinking about core ideas over multiple years. "If mastery of a core idea in a science discipline is the ultimate educational destination, then well-designed learning progressions provide a map of the routes that can be taken to reach that destination" (National Research Council, *A Framework for K–12 Science Education*, 2011). Most of this work is behind the scenes, never seen by the user of the FOSS Program. It does surface, however, in two places: (1) the **conceptual framework** represents the structure of scientific knowledge taught and assessed in a module, and (2) the **conceptual flow** is a graphic and narrative description of the sequence of ideas, presented in the Background for the Teacher section of each investigation.

The FOSS modules are organized into three domains: physical science, earth science, and life science. Each domain is divided into two strands, which represent a core scientific idea, as shown in the columns in the table: matter/energy and change, dynamic atmosphere/rocks and landforms, structure and function/complex systems. The sequence of modules in each strand relates to the core ideas described in the national framework. Modules at the bottom of the table form the foundation in the primary grades. The core ideas develop in complexity as you proceed up the columns.

**FOSS Elementary Module Sequences**

| | PHYSICAL SCIENCE | | EARTH SCIENCE | | LIFE SCIENCE | |
|---|---|---|---|---|---|---|
| | **MATTER** | **ENERGY AND CHANGE** | **DYNAMIC ATMOSPHERE** | **ROCKS AND LANDFORMS** | **STRUCTURE/ FUNCTION** | **COMPLEX SYSTEMS** |
| 6 | Mixtures and Solutions | Motion, Force, and Models | Weather on Earth | Sun, Moon, and Planets | Living Systems | |
| | Measuring Matter | Energy and Electromagnetism | Water | Soils, Rocks, and Landforms | Structures of Life | Environments |
| | Solids and Liquids | Balance and Motion | Air and Weather | Pebbles, Sand, and Silt | Plants and Animals | Insects and Plants |
| K | Materials in Our World | | Trees and Weather | | Animals Two by Two | |

**FOSS**

In addition to the science content framework, every module provides opportunities for students to engage in and understand scientific practices, and many modules explore issues related to engineering practices and the use of natural resources.

**TEACHING NOTE**

*A Framework for K–12 Science Education describes these eight scientific and engineering practices as essential elements of a K–12 science and engineering curriculum.*

## Asking questions and defining problems

- Ask questions about objects, organisms, systems, and events in the natural and human-made world (science).

- Ask questions to define a problem, determine criteria for solutions, and identify constraints (engineering).

## Planning and carrying out investigations

- Plan and conduct investigations in the laboratory and in the field to gather appropriate data (describe procedures, determine observations to record, decide which variables to control) or to gather data essential for specifying and testing engineering designs.

## Analyzing and interpreting data

- Use a range of tools (numbers, words, tables, graphs, images, diagrams, equations) to organize observations (data) in order to identify significant features and patterns.

## Developing and using models

- Use models to help develop explanations, make predictions, and analyze existing systems, and recognize strengths and limitations of the models.

## Using mathematics and computational thinking

- Use mathematics and computation to represent physical variables and their relationships.

## Constructing explanations and designing solutions

- Construct logical explanations of phenomena, or propose solutions that incorporate current understanding or a model that represents it and is consistent with the available evidence.

## Engaging in argument from evidence

- Defend explanations, formulate evidence based on data, examine one's own understanding in light of the evidence offered by others, and collaborate with peers in searching for explanations.

## Obtaining, evaluating, and communicating information

- Communicate ideas and the results of inquiry—orally and in writing—with tables, diagrams, graphs, and equations and in discussions with peers.

TEACHING NOTE

*Crosscutting concepts, as identified by the national framework, bridge disciplinary core ideas and provide an organizational framework for connecting knowledge from different disciplines into a coherent and scientifically based view of the world. The **Trees and Weather Module** addresses these crosscutting concepts: patterns, and stability and change.*

# CONCEPTUAL FRAMEWORK
## *in Trees and Weather*

This module provides foundational experiences with both life science and earth science concepts. In the FOSS Program, this module is part of the earth science strand. It emphasizes key earth science concepts including attributes of landforms and weather, and their effect on life on Earth. Students keep a weather calendar and monitor general weather conditions day by day. They focus on temperature and wind (moving air) as two important aspects of weather. They look at seasonal weather changes and their impact on trees. Students will come away with a respect for trees and other important resources, and know that they should, and can, be conserved.

Trees grow just about everywhere. They are found high in mountains and below sea level in arid, salty deserts. Some trees are adapted to withstand prolonged droughts, while others grow in water. Some are huge, weighing many tons, while others are only a few centimeters tall. Trees are well represented throughout the world, and everyone is familiar with trees of one kind or another from early childhood.

Exactly what makes a plant a tree is not precisely defined. Trees usually have single woody stems, called trunks, that are covered in a tough outer layer called bark. Trees tend to be large, relatively long-lived organisms. But the definition is left up to the subjective determination of an individual observer. When asked to point out a tree, virtually every kindergartner will do so without hesitation and will invariably be right.

For the sake of this module, trees are those big, living plants that grow in the schoolyard, along the streets, and around the homes in your community. They have thick trunks covered with rough bark. Higher up, the trunks give way to branches, and eventually each branch and twig terminates in leaves.

The study of trees is a study of the commonplace. But regard that tree from a fresh point of view, and you've just engaged in one of the delightful aspects of science—the power it has to make a mundane object or event provocative and exciting.

Some trees can be identified by the shape of their silhouettes. The distinctive shapes of the weeping willow and the coconut palm are easy to discern at a glance, but with a little experience, dozens of trees take on a distinctive shape in the eyes of the viewer.

---

### Life Science, Structure and Function:
### Trees and Weather

**Life has structure and function**

**Concept A**  The internal and external structures of plants and animals serve various functions in growth, survival, and reproduction.

**Concept C**  Organisms need matter and energy to live and grow.

Similarly, leaves of every tree have evolved so that they are unique in shape and size for each kind of tree. Many people can spot the characteristic shape of a maple leaf or a pine needle or the dramatic convolutions of an oak leaf. As the student is exposed to a greater array of leaves, they all start to reveal themselves as unique and understandable. Before long, a score of oaks or a dozen maples can be identified.

Although some students will demonstrate an amazing ability to discriminate and identify large numbers of trees, that is not a major goal of this module. These activities are designed to use trees as typical examples of plants. The strategies used to bring students into meaningful interaction with trees would work equally well to teach them about grasses, flowers, seaweeds, or mosses. The strategies encourage students to step up close and look at the fine structure, to step back and look at the gross structure, and to compare parts of one tree to the same parts of another. We want early-childhood students to come away with a developing concept of what makes a plant a tree, and to be able to describe some of the fine structure of trees. If students have opportunities to "adopt" and observe a tree over time, they will begin to understand change through the seasons—to discover what changes, what does not change, and when the changes take place.

In addition to the direct experiences with trees and the opportunities to manipulate representational and symbolic materials to enrich the direct experience, FOSS investigations promote communication of students' observations and perceptions. Students talk, draw, and write to communicate their ideas.

During the development of the activities in this module, we asked students to draw a picture of a tree that would tell us about the tree they had adopted and observed in the schoolyard. We were surprised to see how similar many of the pictures were. They were drawn with massive rectangular trunks and round, green canopies right on top. We asked ourselves whether students were really looking at the trees. But we discovered that if we approached a tree, got low like a primary student, and looked up, we perceived a large rectangle topped with a green circular canopy. It was a delightful revelation to discover that kindergartners could see their tree better than the adults who were hovering around.

---

### Earth Science, Dynamic Atmosphere: Trees and Weather

**Structure of Earth**

**Concept A** The hydrosphere has properties that can be observed and quantified.

- Water exists in three states on Earth: solid, liquid, and gas.

**Concept B** The atmosphere has properties that can be observed and quantified.

- Wind is moving air.

**Concept C** Humans depend on Earth's land, ocean, atmosphere, and biosphere for many different resources.

- People use earth materials to make and construct things.

- People can conserve resources.

**Earth Interactions**

**Concept A** Weather and climate are influenced by interactions of the Sun, the ocean, the atmosphere, ice, landforms, and living things.

- Energy that drives weather comes from the Sun.

- Weather describes the minute-by-minute, day-by-day variation of the atmosphere's condition on a local scale.

- Scientists record weather patterns to make predictions.

---

## Dynamic Atmosphere Content Sequence

This table shows the five FOSS modules and courses that address the content sequence "dynamic atmosphere" for grades K–8. Running through the sequence are the two progressions—structure of Earth and Earth interactions. The supporting elements in each module (somewhat abbreviated) are listed. The elements for the **Trees and Weather Module** are expanded to show how they fit into the sequence.

| | DYNAMIC ATMOSPHERE | |
|---|---|---|
| **Module or course** | **Structure of Earth** | **Earth interactions** |
| **Weather and Water** | • Weather is the condition of Earth's atmosphere at a given time in a local place; climate is the range of an area's weather conditions over years.<br>• Weather happens in the troposphere.<br>• Density is a ratio of a mass and its volume.<br>• The angle at which light from the Sun strikes the surface of Earth is the solar angle. | • Complex patterns of interactions determine local weather patterns.<br>• Energy transfers from one place to another by radiation and conduction.<br>• Convection is the circulation of a fluid that results from energy transfer in a fluid.<br>• When air masses of different densities meet, weather changes.<br>• The Sun's energy drives the water cycle and weather. |
| **Weather on Earth** | • Weather is described in terms of variables including temperature, humidity, wind, and air pressure.<br>• Scientists observe, measure, and record patterns of weather to make predictions.<br>• The Sun is the major source of energy that heats Earth; land, water, and air heat up at different rates.<br>• Most of Earth's water is in the ocean. | • The different energy-absorbing properties of earth materials lead to uneven heating of Earth's surface and convection currents.<br>• Evaporation and condensation contribute to the movement of water through the water cycle.<br>• Climate—the range of an area's typical weather conditions—is changing globally; this change will impact all life. |
| **Water** | • Water is found almost everywhere on Earth, e.g., vapor, clouds, rain, snow, ice.<br>• Water expands when heated, contracts when cooled, and expands when frozen.<br>• Cold water is more dense than warmer water; liquid water is more dense than ice.<br>• Soils retain more water than rock particles alone. | • Water moves downhill; the steeper the slope, the faster water moves.<br>• Ice melts when heated; liquid water freezes when cooled.<br>• Evaporation is the process by which liquid (water) changes into gas (water vapor).<br>• Condensation is the process by which gas (water vapor) changes into liquid (water). |
| **Air and Weather** | • Air is a gas and is all around us.<br>• Air is matter and takes up space.<br>• Weather describes conditions in the air.<br>• Weather conditions can be measured.<br>• Clouds are made of liquid water drops.<br>• Natural sources of water include streams, rivers, lakes, and the ocean. | • The Sun heats Earth during the day.<br>• Compressed air can move things.<br>• Daily changes in weather conditions can be observed, compared, and predicted.<br>• Each season has typical weather conditions.<br>• Weather affects animals and plants. |
| **Trees and Weather** | | |

The **Trees and Weather Module** aligns with the *NRC Framework*. The module addresses these K–2 grade band endpoints described for core ideas from the national framework for **Earth's systems** and **Earth and human activity**.

## Earth and Space Sciences

### Core idea ESS2: Earth's systems—How and why is Earth constantly changing?

- *ESS2.D: What regulates weather and climate?* [Weather is the combination of sunlight, wind, snow or rain, and temperature in a particular region at a particular time. People measure these conditions to describe and record the weather and to notice patterns over time.]

- *ESS2.E: How do living organisms alter Earth's processes and structures?* [Plants and animals (including humans) depend on the land, water, and air to live and grow. They in turn can change their environment.]

### Core idea ESS3: Earth and human activity—How do Earth's surface processes and human activities affect each other?

- *ESS3.A: How do humans depend on Earth's resources?* [Living things need water, air, and resources from the land, and they try to live in places that have the things they need. Humans use natural resources for everything they do: for example, they use soil and water to grow food.]

| Structure of Earth | Earth interactions |
|---|---|
| • Weather is the condition of the air outside; weather changes.<br>• Temperature is how hot or cold it is, and can be measured with a thermometer.<br>• Wind is moving air; wind socks indicate direction and speed. | • Each season has typical weather conditions that can be observed, compared, and predicted.<br>• Trees change through the seasons. |

*Trees and Weather*

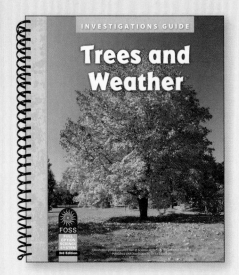

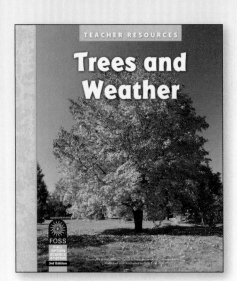

# FOSS COMPONENTS

## Teacher Toolkit

The *Teacher Toolkit* is the most important part of the FOSS Program. It is here that all the wisdom and experience contributed by hundreds of educators has been assembled. Everything we know about the content of the module, how to teach the subject, and the resources that will assist the effort are presented here. Each toolkit has three parts.

***Investigations Guide.*** This spiral-bound document contains these chapters.

- Overview

- Materials

- Investigations (four in this module)

***Teacher Resources.*** This three-ring binder contains these chapters.

- FOSS Introduction

- Assessment

- Science Notebooks in Grades K–2

- Science-Centered Language Development

- Taking FOSS Outdoors

- FOSSweb and Technology

- Science Notebook Masters (for grades 1–6)

- Teacher Masters

- Assessment Masters

The chapters in *Teacher Resources* and the Spanish duplication masters can also be found on FOSSweb (www.FOSSweb.com) and on CDs included in the *Teacher Toolkit.*

***FOSS Science Resources.*** One copy of the student book of readings is included in the *Teacher Toolkit.*

## Equipment Kit

The FOSS Program provides the materials needed for the investigations, including metric measuring tools, in sturdy, front-opening drawer-and-sleeve cabinets. Inside, you will find high-quality materials packaged for a class of 32 students. Consumable materials are supplied for two uses before you need to restock. In addition, you will be asked to supply small quantities of common classroom items.

## FOSS Science Resources Books

*FOSS Science Resources: Trees and Weather* is a book of original readings developed to accompany this module. The readings are referred to as articles in the *Investigations Guide*. Students read the articles in the book as they progress through the module. The articles cover a specific concept, usually after that concept has been introduced in an active investigation.

The articles in *Science Resources* and the discussion questions provided in the *Investigations Guide* help students make connections to the science concepts introduced and explored during the active investigations. Concept development is most effective when students experience organisms, objects, and phenomena firsthand before engaging the concepts in text. The text and illustrations help make connections between what students experience concretely and the ideas that explain their observations.

## FOSSweb and Technology

The FOSS website opens new horizons for educators, students, and families, in the classroom or at home. Each module has an interactive site where students and families can find instructional activities, interactive simulations and virtual investigations, and other resources. FOSSweb provides resources for materials management, general teaching tools for FOSS, purchasing links, contact information for the FOSS Project, and technical support. You do not need an account to view this general FOSS Program information. In addition to the general information, FOSSweb provides digital access to PDF versions of the *Teacher Resources* component of the *Teacher Toolkit* and digital-only resources that supplement the print and kit materials.

Additional resources are available to support FOSS teachers. With an educator account, you can customize your homepage, set up easy access to the digital components of the modules you teach, and create class pages for your students with access to tutorials and online assessments.

## Ongoing Professional Development

The Lawrence Hall of Science and Delta Education are committed to supporting science educators with unrivaled teacher support, high-quality implementation, and continuous staff-development opportunities and resources. FOSS has a strong network of consultants who have rich and experienced backgrounds in diverse educational settings using FOSS. Find out about professional-development opportunities on FOSSweb.

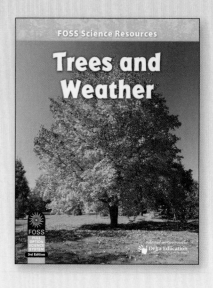

▶ **NOTE**

*FOSS Science Resources: Trees and Weather* is also provided as a big book in the kit.

▶ **NOTE**

To access all the teacher resources and to set up customized pages for using FOSS, log in to FOSSweb through an educator account.

# FOSS INSTRUCTIONAL DESIGN

Each FOSS investigation follows a similar design to provide multiple exposures to science concepts. The design includes these pedagogies.

- Active investigation, including outdoor experiences
- Recording in science notebooks to answer the focus question
- Reading in *FOSS Science Resources*
- Assessment to monitor progress and motivate student reflection on learning

In practice, these components are seamlessly integrated into a continuum designed to maximize every student's opportunity to learn. An instructional sequence may move from one pedagogy to another and back again to ensure adequate coverage of a concept.

## FOSS Investigation Organization

Modules are subdivided into **investigations** (four in this module). Investigations are further subdivided into **parts**. Each part of each investigation is driven by a **focus question**. The focus question, usually presented as the part begins, signals the challenge to be met, mystery to be solved, or principle to be uncovered. The focus question guides students' actions and thinking and makes the learning goal of each part explicit for teachers. Each part concludes with students recording an answer to the focus question in their notebooks.

Investigation-specific **scientific background** information for the teacher is presented in each investigation chapter. The content discussion is divided into sections, each of which relates directly to one of the focus questions. This section ends with information about teaching and learning and a conceptual-flow diagram for the content.

The **Getting Ready** and **Guiding the Investigation** sections have several features that are flagged or presented in the sidebars. These include several icons to remind you when a particular pedagogical method is suggested, as well as concise bits of information in several categories.

**Teaching notes** appear in blue boxes in the sidebars. These notes comprise a second voice in the curriculum—an educative element. The first (traditional) voice is the message you deliver to students. It supports your work teaching students at all levels, from management to inquiry. The second educative voice, shared as a teaching note, is designed to help you understand the science content and pedagogical rationale at work behind the instructional scene.

**FOCUS QUESTION**

*What are the parts of trees?*

**TEACHING NOTE**

*This focus question can be answered with a simple yes or no, but the question has power when students support their answers with evidence. Their answers should take the form "Yes, because _____ ."*

The **safety** icon alerts you to a potential safety issue. It could relate to the use of a chemical substance, such as salt, requiring safety goggles, or the possibility of a student allergic reaction when students use latex, legumes, or wheat.

The small-group **discussion** icon asks you to pause while students discuss data or construct explanations in their groups.

The **new-word** icon alerts you to a new vocabulary word or phrase that should be introduced thoughtfully. The new vocabulary should also be entered onto the word wall (or pocket chart). A complete list of the scientific vocabulary used in each investigation appears in the sidebar on the last page of the Background for the Teacher section.

The **vocabulary** icon indicates where students should review recently introduced vocabulary, often just before they will be answering the focus question or preparing for benchmark assessment.

The **recording** icon points out where students should make a science-notebook entry. Students record on prepared notebook sheets or, increasingly, on pages in their science notebooks.

The **reading** icon signals when the class should read a specific article in the *FOSS Science Resources* book, preferably during a reading period.

The **assessment** icon appears when there is an opportunity to assess student progress, by using embedded assessments. Some of the embedded-assessment methods for grades K–2 include observation of students engaged in scientific practices and review of a notebook entry (drawing or text).

The **outdoor** icon signals when to move the science learning experience into the schoolyard. It also helps you plan for selecting and preparing an outdoor site for a student activity.

The **engineering** icon indicates opportunities for addressing engineering practices—applying and using scientific knowledge. These opportunities include developing a solution to a problem, constructing and evaluating models, and using systems thinking.

The **EL note** in the sidebar provides a specific strategy to assist English learners in developing science concepts. A discussion of strategies is in the Science-Centered Language Development chapter.

### EL NOTE

*See the Science-Centered Language Development chapter for notebook-sharing strategies.*

To help with pacing, you will see icons for **breakpoints**. Some breakpoints are essential, and others are optional.

## POSSIBLE BREAKPOINT

## Active Investigation

Active investigation is a master pedagogy. Embedded within active learning are a number of pedagogical elements and practices that keep active investigation vigorous and productive. The enterprise of active investigation includes

- context: questioning and planning;
- activity: doing and observing;
- data management: recording, organizing, and processing;
- analysis: discussing and writing explanations.

***Context: questioning and planning.*** Active investigation requires focus. The context of an inquiry can be established with a focus question or challenge from you or, in some cases, from students. (What shapes are leaves?) At other times, students are asked to plan a method for investigation. This might start with a teacher demonstration or presentation. Then you challenge students to plan an investigation, such as to find out what a wind sock tells us about the wind. In either case, the field available for thought and interaction is limited. This clarification of context and purpose results in a more productive investigation.

***Activity: doing and observing.*** In the practice of science, scientists put things together and take things apart, observe systems and interactions, and conduct experiments. This is the core of science—active, firsthand experience with objects, organisms, materials, and systems in the natural world. In FOSS, students engage in the same processes. Students often conduct investigations in collaborative groups of four, with each student taking a role to contribute to the effort.

The active investigations in FOSS are cohesive, and build on each other and the readings to lead students to a comprehensive understanding of concepts. Through the investigations, students gather meaningful data.

***Data management: recording, organizing, and processing.*** Data accrue from observation, both direct (through the senses) and indirect (mediated by instrumentation). Data are the raw material from which scientific knowledge and meaning are synthesized. During and after work with materials, students record data in their science notebooks. Data recording is the first of several kinds of student writing.

Students then organize data so that they will be easier to think about. Tables allow efficient comparison. Organizing data in a sequence (time) or series (size) can reveal patterns. Students process some data into graphs, providing visual display of numerical data. They also organize data and process them in the science notebook.

***Analysis: discussing and writing explanations.*** The most important part of an active investigation is extracting its meaning. This constructive process involves logic, discourse, and existing knowledge. Students share their explanations for phenomena, using evidence generated during the investigation to support their ideas. Students conclude the active investigation by writing a summary of their learning in their science notebooks as well as questions raised during the activity.

## Science Notebooks

Research and best practice have led FOSS to place more emphasis on the student science notebook. Keeping a notebook helps students organize their observations and data, process their data, and maintain a record of their learning for future reference. The process of writing about their science experiences and communicating their thinking is a powerful learning device for students. The science-notebook entries stand as credible and useful expressions of learning. The artifacts in the notebooks form one of the core elements of the assessment system.

## Reading in *FOSS Science Resources*

The *FOSS Science Resources* books emphasize expository articles and biographical sketches. FOSS suggests that the reading be completed during language-arts time. When language-arts skills and methods are embedded in content material that relates to the authentic experience students have had during the FOSS active learning sessions, students are interested, and they get more meaning from the text material.

## Assessing Progress for Kindergarten

Assessment and teaching must be woven together to provide the greatest benefit to both the student and the teacher. Assessing young students is a process of planning what to assess, and observing, questioning, and recording information about student learning for future reference. Observing students as they engage in the activity and as they share notebook entries (drawings and words) reveals their thinking and problem-solving abilities. Questioning probes for understanding. Both observing and questioning will give you information about what individual students can and can't do, and what they know or don't know. This information allows you to plan your instruction thoughtfully. For example, if you find students need more experience comparing the properties of leaves, you can provide more time at a center that focuses on comparing similar items or select extension activities that will continue to develop the ability to identify similarities and differences.

Use the techniques that work for you and your students and that fit with the overall kindergarten curriculum goals. The most detailed and reliable picture of students' growth emerges from information gathered by a variety of assessment strategies.

FOSS embedded assessments for kindergarten allow you and your students to monitor learning on a daily basis as you progress through the **Trees and Weather Module**. You will find suggestions for what to assess in the Getting Ready section of each part of each investigation. For example, here is the Getting Ready step for Part 1 of the first investigation.

## 14. Plan assessment for Part 1

There are six objectives that can be assessed at any time during any part of this investigation.

### What to Look For

- *Students ask questions.*

- *Students use their senses to observe living things.*

- *Students show respect for living things.*

- *Students record observations.*

- *Students communicate observations orally, in writing, and in drawings.*

- *Students use new vocabulary.*

Here are specific content objectives for this part.

- *Trees are living plants; trees have basic needs.*

- *Trees have structures.*

- *Trees are a natural resource.*

Focus on a few students each session. Record the date and a + or − on the *Assessment Checklist*.

Make copies of the *Assessment Checklist*, attach them to a clipboard, and carry them with you when students are engaged in the investigations. Record your observations as you interact with students, or take a few minutes after class to reflect on the lesson.

**TEACHING NOTE**

*Because there are several opportunities for you to assess students on each objective, we suggest that you focus on six to ten students during each session rather than trying to assess the whole class at one time.*

## Taking FOSS Outdoors

FOSS throws open the classroom door and proclaims the entire school campus to be the science classroom. The true value of science knowledge is its usefulness in the real world and not just in the classroom. Taking regular excursions into the immediate outdoor environment has many benefits. First of all, it provides opportunities for students to apply things they learned in the classroom to novel situations. When students are able to transfer knowledge of scientific principles to natural systems, they experience a sense of accomplishment.

In addition to transfer and application, students can learn things outdoors that they are not able to learn indoors. The most important object of inquiry outdoors is the outdoors itself. To today's youth, the outdoors is something to pass through as quickly as possible to get to the next human-managed place. For many, engagement with the outdoors and natural systems must be intentional, at least at first. With repeated visits to familiar outdoor learning environments, students may first develop comfort in the outdoors, and then a desire to embrace and understand natural systems.

Most investigations include an outdoor experience. Venturing out will require courage the first time or two you mount an outdoor expedition. It will confuse students as they struggle to find the right behavior that is a compromise between classroom rigor and diligence and the freedom of recreation. With persistence, you will reap rewards. You will be pleased to see students' comportment develop into proper field-study habits, and you might be amazed by the transformation of students who have behavior issues in the classroom but who become insightful observers and leaders in the schoolyard environment.

Teaching outdoors is the same as teaching indoors—except for the space. You need to manage the same four core elements of teaching: time, space, materials, and students. Because of the different space, new management procedures are required. Students can get farther away. Materials have to be transported. The space has to be defined and honored. Time has to be budgeted for getting to, moving around in, and returning from the outdoor study site. All these and more issues and solutions are discussed in the Taking FOSS Outdoors chapter in *Teacher Resources*.

FOSS is very enthusiastic about this dimension of the program and looks forward to hearing about your experience using the schoolyard as a logical extension of your classroom.

# Science-Centered Language Development

The FOSS active investigations, science notebooks, *FOSS Science Resources* articles, and formative assessments provide rich contexts in which students develop and exercise thinking and communication. These elements are essential for effective instruction in both science and language arts—students experience the natural world in real and authentic ways and use language to inquire, process information, and communicate their thinking about scientific phenomena. FOSS refers to this development of language process and skills within the context of science as science-centered language development.

In the Science-Centered Language Development chapter in *Teacher Resources*, we explore the intersection of science and language and the implications for effective science teaching and language development. We identify best practices in language-arts instruction that support science learning and examine how learning science content and engaging in scientific practices support language development.

Language plays two crucial roles in science learning: (1) it facilitates the communication of conceptual and procedural knowledge, questions, and propositions, and (2) it mediates thinking—a process necessary for understanding. For students, language development is intimately involved in their learning about the natural world. Science provides a real and engaging context for developing literacy, and language-arts skills and strategies support conceptual development and scientific practices. For example, the skills and strategies used for enhancing reading comprehension, writing expository text, and exercising oral discourse are applied when students are recording their observations, making sense of science content, and communicating their ideas. Students' use of language improves when they discuss (speak and listen, as in the Wrap-Up/Warm-Up activities), write, and read about the concepts explored in each investigation.

There are many ways to integrate language into science investigations. The most effective integration depends on the type of investigation, the experience of students, the language skills and needs of students, and the language objectives that you deem important at the time. The Science-Centered Language Development chapter is a library of resources and strategies for you to use. The chapter describes how literacy strategies are integrated purposefully into the FOSS investigations, gives suggestions for additional literacy strategies that both enhance students' learning in science and develop or exercise English-language literacy skills, and develops science vocabulary with scaffolding strategies for supporting all learners. The last section covers language-development strategies specifically for English learners.

> **TEACHING NOTE**
>
> *Embedded even deeper in the FOSS pedagogical practice is a bolder philosophical stance. Because language arts commands the greatest amount of the instructional day's time, FOSS has devoted a lot of creative energy to defining and exploring the relationship between science learning and the development of language-arts skills. FOSS elucidates its position in the Science-Centered Language Development chapter.*

# FOSSWEB AND TECHNOLOGY

FOSS is committed to providing a rich, accessible technology experience for all FOSS users. FOSSweb is the Internet access to FOSS digital resources. It provides enrichment for students and support for teachers, administrators, and families who are actively involved in implementing and enjoying FOSS materials. Here are brief descriptions of selected resources to help you get started with FOSS technology.

## Technology to Engage Students at School and at Home

**Multimedia activities.** The multimedia simulations and activities were designed to support students' learning. They include virtual investigations and student tutorials that you can use to support students who have difficulties with the materials or who have been absent.

**FOSS Science Resources.** The student reading book is available as an audio book on FOSSweb, accessible at school or at home. In addition, as premium content, *FOSS Science Resources* is available as an eBook. The eBook supports a range of font sizes and can be projected for guided reading with the whole class as needed.

**Home/school connection.** Each module includes a letter to families, providing an overview of the goals and objectives of the module. Most investigations have a home/school activity that provides science experiences to connect the classroom experiences with students' lives outside of school. These connections are available in print in the *Teacher Resources* binder and on FOSSweb.

**Student media library.** A variety of media enhance students' learning. Formats include photos, videos, an audio version of each student book, and frequently asked science questions. These resources are also available to students when they log in with a student account.

**Recommended books and websites.** FOSS has reviewed print books and digital resources that are appropriate for students and prepared a list of these media resources.

**Class pages.** Teachers with a FOSSweb account can easily set up class pages with notes and assignments for each class. Students and families can then access this class information online.

▶ **NOTE**
The FOSS digital resources are available online at FOSSweb. You can always access the most up-to-date technology information, including help and troubleshooting, on FOSSweb. See the FOSSweb and Technology chapter for a complete list of these resources.

## Technology to Support Teachers

**Teacher-preparation video.** The video presents information to help you prepare for a module, including detailed investigation information, equipment setup and use, safety, and what students do and learn through each part of the investigation.

**Science-notebook masters and teacher masters.** All notebook masters (grades 1–6) and teacher masters used in the modules are available digitally on FOSSweb for downloading and for projection during class. These sheets are available in English and Spanish.

**Focus questions.** The focus questions for each investigation are formatted for classroom projection and for printing onto labels that students can glue into their science notebooks.

**Equipment photo cards.** The cards provide labeled photos of equipment supplied in each FOSS kit.

**Materials Safety Data Sheets (MSDS).** These sheets have information from materials manufacturers on handling and disposal of materials.

**Teacher Resources chapters.** FOSSweb provides PDF files of all chapters from the *Teacher Resources* binder.

- Assessment

- Science Notebooks

- Science-Centered Language Development

- Taking FOSS Outdoors

- FOSSweb and Technology

**Streaming video.** Some video clips are part of the instruction in the investigation, and others extend concepts presented in a module.

**Resources by investigation.** This digital listing provides online links to notebook sheets, assessment and teacher masters, and multimedia for each investigation of a module, for projection in the classroom.

**Interactive-whiteboard resources.** You can use these slide shows and other resources with an interactive whiteboard.

**Investigations eGuide.** The eGuide is the complete FOSS *Investigations Guide* component of the *Teacher Toolkit* in an electronic web-based format, allowing access from any Internet-enabled computer.

> **NOTE**
> The Spanish masters are available only on FOSSweb and on one of the CDs provided in the *Teacher Toolkit*.

# UNIVERSAL DESIGN FOR LEARNING

The roots of FOSS extend back to the mid-1970s and the Science Activities for the Visually Impaired and Science Enrichment for Learners with Physical Handicaps projects (SAVI/SELPH). As those special-education science programs expanded into fully integrated settings in the 1980s, hands-on science proved to be a powerful medium for bringing all students together. The subject matter is universally interesting, and the joy and satisfaction of discovery are shared by everyone. Active science by itself provides part of the solution to full inclusion.

Many years later, FOSS began a collaboration with educators and researchers at the Center for Applied Special Technology (CAST), where principles of Universal Design for Learning (UDL) had been developed and applied. FOSS continues to learn from our colleagues about ways to use new media and technologies to improve instruction. Here are the UDL principles.

Principle 1. Provide multiple means of representation. Give learners various ways to acquire information and knowledge.

Principle 2. Provide multiple means of action and expression. Offer students alternatives for demonstrating what they know.

Principle 3. Provide multiple means of engagement. Help learners get interested, be challenged, and stay motivated.

The FOSS Program has been designed to maximize the science-learning opportunities for students with special needs and students from culturally and linguistically diverse origins. FOSS is rooted in a 30-year tradition of multisensory science education and informed by recent research on UDL. Strategies found effective with students with special needs and students who are learning English are incorporated into the materials and procedures used with all students.

## English Learners

The FOSS multisensory program provides a rich laboratory for language development for English learners. The program uses a variety of techniques to make science concepts clear and concrete, including modeling, visuals, and active investigations in small groups at centers. Key vocabulary is usually developed within an activity context with frequent opportunities for interaction and discussion between teacher and student and among students. This provides practice and application of the new vocabulary. Instruction is guided and scaffolded through carefully designed lesson plans, and students are supported throughout. The learning is active and engaging for all students, including English learners.

Science vocabulary is introduced in authentic contexts while students engage in active learning. Strategies for helping all primary students read, write, speak, and listen are described in the Science-Centered Language Development chapter. There is a section on science-vocabulary development with scaffolding strategies for supporting English learners. These strategies are essential for English learners, and they are good teaching strategies for all learners.

## Differentiated Instruction

FOSS instruction allows students to express their understanding through a variety of modalities. Each student has multiple opportunities to demonstrate his or her strengths and needs. The challenge is then to provide appropriate follow-up experiences for each student. For some students, appropriate experience might mean more time with the active investigations. For other students, it might mean more experience building explanations of the science concepts orally or in writing or drawing. For some students, it might mean making vocabulary more explicit through new concrete experiences or through reading to students. For some students, it may be scaffolding their thinking through graphic organizers. For other students, it might be designing individual projects or small-group investigations. For some students, it might be more opportunities for experiencing science outside the classroom in more natural, outdoor environments.

There are several possible strategies for providing differentiated instruction. The FOSS Program provides tools and strategies so that you know what students are thinking throughout the module. Based on that knowledge, read through the extension activities for experiences that might be appropriate for students who need additional practice with the basic concepts as well as those ready for more advanced projects. Interdisciplinary extensions are listed at the end of each investigation. Use these ideas to meet the individual needs and interests of your students.

# ORGANIZING THE CLASSROOM

Students in primary grades are usually most comfortable working as individuals with materials. The abilities to share, take turns, and learn by contributing to a group goal are developing but are not reliable as learning strategies all the time. Because of this egocentrism and the need for many students to control materials or dominate actions, the FOSS kit includes a lot of materials. To effectively manage students and materials, FOSS offers some suggestions.

## Small-Group Centers

Many of the kindergarten-level observations and investigations are conducted with small groups at a learning center. Limit the number of students at the center to six to ten at one time. When possible, each student will have his or her own equipment to work with. In some cases, students will have to share materials and equipment and make observations together. Primary students are good at working together independently.

As one group at a time is working at the center on a FOSS activity, other students will be doing something else. Over the course of an hour or more, plan to rotate all students through the center, or allow the center to be a free-choice station.

## Whole-Class Activities

Introducing and wrapping up the center activities require you to work for brief periods with the whole class. FOSS suggests for these introductions and wrap-ups that you gather the class at the rug or other location in the classroom where students can sit comfortably in a large group.

## Guides for Adult Helpers

In the *Teacher Resources* binder, you will find center instructions sheet duplication masters for some investigation parts. These sheets are intended as a quick reference for a family member or other adult who might be supervising the center. The sheets help that person keep the activity moving in a productive direction. The sheets can be laminated or slipped into a clear plastic sheet protector for durability.

## When You Don't Have Adult Helpers

Some parts of investigations are designed for small groups, with an aide or a student's family member available to guide the activity and to encourage discussion and vocabulary development. We realize that there are many primary classrooms in which the teacher is the only adult present. Here are some ways to manage in that situation.

- Invite upper-elementary students to visit your class to help with the activities. They should be able to read the center instructions sheets and conduct the activities with students. Remind older students to be guides and to let primary students do the activities themselves.

- Introduce each part of the activity with the whole class. Set up the center as described in the *Investigations Guide*, but let students work at the center by themselves. Discussion may not be as rich, but most of the centers can be done independently by students once they have been introduced to the process. Be a 1-minute manager, checking on the center from time to time, offering a few words of advice or direction.

## When Students Are Absent

When a student is absent for an activity, give him or her a chance to spend some time with the materials at a center. Another student might act as a peer tutor. Allow the student to bring home a *FOSS Science Resources* book to read with a family member.

# SAFETY IN THE CLASSROOM AND OUTDOORS

Following the procedures described in each investigation will make for a very safe experience in the classroom. You should also review your district safety guidelines and make sure that everything you do is consistent with those guidelines. Two posters are included in the kit: *Science Safety* for classroom use and *Outdoor Safety* for outdoor activities.

Look for the safety icon in the Getting Ready and Guiding the Investigation sections that will alert you to safety considerations throughout the module.

Materials Safety Data Sheets (MSDS) for materials used in the FOSS Program can be found on FOSSweb. If you have questions regarding any MSDS, call Delta Education at 1-800-258-1302 (Monday–Friday, 8 a.m.–6 p.m. EST).

## Science Safety in the Classroom

General classroom safety rules to share with students are listed here.

1. Listen carefully to your teacher's instructions. Follow all directions. Ask questions if you don't know what to do.

2. Tell your teacher if you have any allergies.

3. Never put any materials in your mouth. Do not taste anything unless your teacher tells you to do so.

4. Never smell any unknown material. If your teacher tells you to smell something, wave your hand over the material to bring the smell toward your nose.

5. Do not touch your face, mouth, ears, eyes, or nose while working with chemicals, plants, or animals.

6. Always protect your eyes. Wear safety goggles when necessary. Tell your teacher if you wear contact lenses.

7. Always wash your hands with soap and warm water after handling chemicals, plants, or animals.

8. Never mix any chemicals unless your teacher tells you to do so.

9. Report all spills, accidents, and injuries to your teacher.

10. Treat animals with respect, caution, and consideration.

11. Clean up your work space after each investigation.

12. Act responsibly during all science activities.

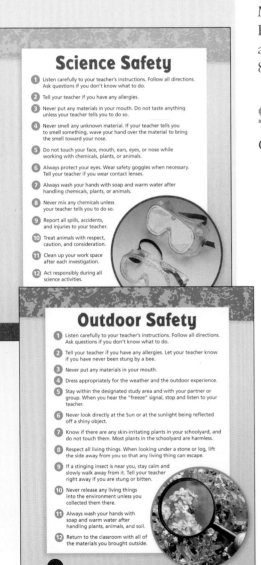

# SCHEDULING THE MODULE

The Getting Ready section for each part of the investigation helps you prepare. It provides information on scheduling the investigation and introduces the tools and techniques used in the investigation. The first item in the Getting Ready section gives an estimated amount of time the part should take. A general rule of thumb is to plan 10 minutes to introduce the investigation to the whole class, about 20–25 minutes at the center for each group, about 10 minutes to wrap up the activity with the whole class, and a few minutes to transition to the groups. Notebook sessions can be done with the whole class after everyone has participated in the center activities. All of the outdoor sessions are whole-class activities.

Below is a list of the investigations and parts and the format of the investigation (whole class, center, or a combination of the two).

▶ **NOTE**
The investigations are numbered, and we suggest that they be conducted in order since the concepts build from investigation to investigation.

Investigation 4 looks at seasonal changes and calls for scheduling activities throughout the year.

| INVESTIGATION | PART | ORGANIZATION |
|---|---|---|
| 1. Observing Trees | 1. Observing Schoolyard Trees | whole class |
| | 2. Tree Parts | whole class/center |
| | 3. Tree Puzzles | center |
| | 4. Tree-Silhouette Cards | whole class/center |
| | 5. Adopt Schoolyard Trees | whole class |
| | 6. A Tree Comes to Class | whole class |
| 2. Observing Leaves | 1. Leaf Walk | whole class |
| | 2. Leaf Shapes | center |
| | 3. Comparing Leaves | whole class |
| | 4. Matching Leaf Silhouettes | center |
| | 5. Leaf Books | center |
| 3. Observing Weather | 1. Weather Calendar | whole class |
| | 2. Recording Temperature | whole class |
| | 3. Wind Direction | whole class/center |
| 4. Trees through the Seasons | 1. Fall: What Comes from Trees? | whole class/center |
| | 2. Fall: Food from Trees | center |
| | 3. Fall: Visiting Adopted Trees | whole class |
| | 4. Winter: Evergreen Hunt | whole class |
| | 5. Winter: Twigs | center |
| | 6. Winter: Visiting Adopted Trees | whole class |
| | 7. Spring: Forcing Twigs | center |
| | 8. Spring: Bark Hunt | whole class |
| | 9. Spring: Visiting Adopted Trees | whole class |

# FOSS K–8 SCOPE AND SEQUENCE

| Grade | Physical Science | Earth Science | Life Science |
|---|---|---|---|
| 6–8 | Electronics<br>Chemical Interactions<br>Force and Motion | Planetary Science<br>Earth History<br>Weather and Water | Human Brain and Senses<br>Populations and Ecosystems<br>Diversity of Life |
| 4–6 | Mixtures and Solutions<br>Motion, Force, and Models<br>Energy and Electromagnetism | Weather on Earth<br>Sun, Moon, and Planets<br>Soils, Rocks, and Landforms | Living Systems<br>Environments |
| 3 | Measuring Matter | Water | Structures of Life |
| 1–2 | Balance and Motion<br>Solids and Liquids | Air and Weather<br>Pebbles, Sand, and Silt | Insects and Plants<br>Plants and Animals |
| K | Materials in Our World | Trees and Weather | Animals Two by Two |

## Contents

Introduction ......................... **29**

Kit Inventory List .................. **30**

Materials Supplied by the
Teacher ............................... **32**

Preparing a New Kit ............. **34**

Preparing the Kit for Your
Classroom ........................... **36**

Care, Reuse, and Recycling .... **39**

# INTRODUCTION

The Trees and Weather kit contains

- *Teacher Toolkit: Trees and Weather*

  1  *Investigations Guide: Trees and Weather*

  1  *Teacher Resources: Trees and Weather*

  1  *FOSS Science Resources: Trees and Weather*

- *FOSS Science Resources: Trees and Weather*
  (1 big book and class set of student books)

- Equipment for 32 students

A new kit contains enough consumable items for at least two classroom uses before you need to resupply. Some of the FOSS early-childhood investigations take place at a science center for groups of six to ten students at a time. For whole-class activities, use a materials station for the class materials.

Individual photos of each piece of FOSS equipment are available online for printing. For updates to information on materials used in this module and access to the Materials Safety Data Sheets (MSDS), go to www.FOSSweb.com. Links to replacement-part lists and customer service are also available on FOSSweb.

▶ **NOTE**
Delta Education Customer Service can be reached at 1-800-258-1302.

# KIT INVENTORY *List*

## Drawer 1—print materials

| | |
|---|---|
| 1 | *Teacher Toolkit: Trees and Weather* (1 *Investigations Guide*, 1 *Teacher Resources*, and 1 *FOSS Science Resources: Trees and Weather*) |
| 32 | *FOSS Science Resources: Trees and Weather*, student books |
| 1 | *FOSS Science Resources: Trees and Weather*, big book |

## Drawer 2—permanent equipment

Equipment Condition

| | | |
|---|---|---|
| 1 | Book, *How Do We Learn?* | |
| 1 | Book, *Our Very Own Tree* | |
| 1 | Book, *Trees* | |
| 3 | Calendars, laminated | |
| 2 | Card sets, Landforms, 24 cards/set | |
| 8 | Card sets, Tree Parts, 15 cards/set | |
| 10 | Fasteners, round hook-and-loop, extras for replacement | |
| 2 | Geometric shapes sets, yellow felt, 6/set | |
| 24 | Glue sticks ✪ | |
| 1 | Hole punch | |
| 4 | Leaf silhouettes sets, green felt, 9/set | |
| 5 | Leaf silhouettes sets, big and little, 6 cards and 1 strip/set | |
| 5 | Leaf silhouettes sets, same size, 6 cards and 1 strip/set | |
| 5 | Leaf silhouettes and outlines sets, 12/set | |
| 2 | Loupes/magnifying lenses | |
| 1 | Poster, *Red Oak*, with library pocket and 1 set of labels | |
| 1 | Poster, *White Pine*, with library pocket and 1 set of labels | |
| 2 | Poster labels, sets, extras for replacement, 5/set | |
| 1 | Poster set, *A Tree Comes to Class*, 4/set | |
| 2 | Posters, *Science Safety* and *Outdoor Safety* | |

✪ These items might occasionally need replacement.

| | | Equipment Condition |
|---|---|---|
| 1 | Puzzle, apple tree, 6 pieces, with reference sheet | |
| 1 | Puzzle, cottonwood tree, 6 pieces, with reference sheet | |
| 1 | Puzzle, cottonwood tree, 9 pieces, with reference sheet | |
| 1 | Puzzle, fir tree, 6 pieces, with reference sheet | |
| 1 | Puzzle, maple tree, 6 pieces, with reference sheet | |
| 1 | Puzzle, oak tree, 6 pieces, with reference sheet | |
| 1 | Puzzle, palm tree, 6 pieces, with reference sheet | |
| 1 | Puzzle, pine tree, 6 pieces, with reference sheet | |
| 1 | Puzzle, pine tree, 9 pieces, with reference sheet | |
| 1 | Puzzle, poplar tree, 6 pieces, with reference sheet | |
| 10 | Puzzle frames, plastic, clear | |
| 1 | Thermometer, FOSS demonstration | |
| 1 | Thermometer, garden | |
| 2 | Tree-trunk rounds (different trees) | |

## Drawer 2—consumable equipment

| | | |
|---|---|---|
| 3 | Crepe paper, rolls, blue, 4.4 cm × 24.6 m (1.75" × 81') | |
| 2 | Self-stick notes, pads, 100/pad | |
| 1 | String, ball, 30 m (100') | |
| 1 | Yarn skein, yellow, 55 m (60 yd.) | |
| 100 | Zip bags, 1 L (1 qt.) | |

# MATERIALS *Supplied by the Teacher*

Each part of each investigation has a Materials section that describes the materials required for that part. It lists materials needed for each student or group of students and for the class.

Be aware that you must supply some items. Each of these items is indicated in the materials list for each part of the investigation with an asterisk (★). Here is a summary list of those items by investigation.

### For most investigations

- 1 Camera
- • Chart paper and marking pen
- • Drawing utensils (crayons, pencils, colored pencils, marking pens)
- • Glue sticks (if you need additional)
- • Glue, white
- • Paper, white
- 1 Pen, marking, permanent, black
- • Science notebooks (composition books)
- 1 Scissors
- • Tape, transparent

### For outdoor investigations

- • Collecting bags for carrying materials
- • Clipboards (optional)

### Investigation 1: Observing Trees

- 2 Cardboard pieces, 31 × 46 cm (12.25" × 18.25")
- 1 Clipboard
- 20 Paper, construction, pieces, different colors 30 × 45 cm (12" × 18")
- • Paper, contact
- • Phone books or catalogs
- • Planting tools (shovel, hose, bucket)
- 1 Tree in a container
- • Water
- • Yarn, cord, or ribbon

## Investigation 2: Observing Leaves

2–3 Basins or boxes for leaves

1 Cardboard box, small, or shopping bag

1 Felt board

• Leaves from trees

16 Paper, white, pieces

5 Sheet protectors, clear-plastic (optional)

## Investigation 3: Observing Weather

2 Basins

5–6 Cups or envelopes

• Paper, construction, fadeless: orange, yellow, green, blue, purple

• Paper, construction, pieces, 10 × 23 cm (4" × 9")

1 Pen, watercolor or overhead transparency

• Tape, masking

1 Ruler

• Water, warm and cold

## Investigation 4: Trees through the Seasons

32 Bark photos

• Bottles, soft-drink, 2 L

• Evergreen leaves, needles, and scales

• Fruits, edible

• Gravel or pebbles

1 Knife, sharp

1 Laminator (optional)

8 Paper, drawing, pieces, 11 × 14 cm (4.25" × 5.5")

• Paper towels

1 Bag for seeds, plastic, 1 L

1 Pruning shears (optional)

• Twigs

• Water

# PREPARING *a New Kit*

If you choose to prepare the materials all at once with a group of volunteers, you can use the following guidelines for organization.

1. **Prepare the center instruction sheets**

   Each investigation part that involves a group of students at a center has a center instruction sheet written for a parent or other adult helper working with students. The sheet summarizes the information provided to the teacher in the *Investigations Guide*. Use the teacher masters to make a copy of each of the center instructions, and either laminate the sheet or put them in clear-plastic sheet protectors. Take time to orient your adult volunteers or aides to the overall purposes of the activities and encourage them to facilitate but not direct student learning at the center. Below are the teacher master numbers for the center instruction sheets.

   No. 4   Center Instructions—Tree-Part Cards

   No. 6   Center Instructions—Tree Puzzles

   No. 7   Center Instructions—Tree-Silhouette Cards

   No. 14 Center Instructions—Leaf Shapes

   No. 19 Center Instructions—Matching Leaf Silhouettes A

   No. 20 Center Instructions—Matching Leaf Silhouettes B

   No. 25 Center Instructions—Wind Direction

   No. 27 Center Instructions—Food from Trees

   No. 28 Center Instructions—Winter Twigs

   No. 29 Center Instructions—Forcing Twigs

2. **Prepare tree-silhouette cards**

   Use teacher master 8 to make copies of the Tree-Silhouette cards. Each student will need one set of the eight cards. Cut all the cards apart on the dotted lines. Store each set in a zip bag.

## 3. Prepare the demonstration thermometer

The demonstration thermometer has both Celsius and Fahrenheit scales. Decide which scale you want to use.

Set up the demonstration thermometer with five color-coded temperature ranges. Students will use the colored areas on the thermometer to help them associate temperatures with how the air feels.

Use fadeless art paper or construction paper to code the temperature ranges listed below. Write the words in black permanent marker. You may want to cover the paper code with clear contact paper. Be careful not to cover the red and white strip that moves up and down in order to change the temperature reading.

Code 30°C–50°C (80°F–120°F) orange. Label it "Hot."
Code 20°C–30°C (65°F–80°F) yellow. Label it "Warm."
Code 10°C–20°C (50°F–65°F) green. Label it "Cool."
Code 0°C–10°C (32°F–50°F) blue. Label it "Cold."
Code the area below 0°C (32°F) purple. Label it "Freezing."

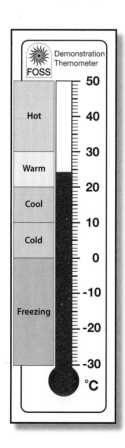

# PREPARING *the Kit for* *Your Classroom*

Some preparation is required each time you use the kit. Doing these things before beginning the module will make daily setup quicker and easier.

1. **Inventory materials**

   Before using a kit, conduct a quick inventory of all items in the kit. You can use the Kit Inventory List provided in this chapter to keep track of any items that are missing or in need of replacement. Information on ordering replacement items can be found at the end of this chapter. The kit contains enough consumables for at least two classes of 32 students.

2. **Use masters to replace labels and cards**

   *Teacher Resources* includes masters for replacing labels and cards that are provided in the kit as equipment. If you find that any of the labels for posters, tree-part cards, or silhouette outlines or cards are missing, you can make your own replacements using these masters.

   > No. 30 Labels for Tree Posters

   > No. 31 Tree-Part Card Masters A

   > No. 32 Tree-Part Card Masters B

   > No. 33 Silhouette Replacements

   > No. 34 Outlines Replacements

   > No. 35 Big and Little Silhouette Replacements A

   > No. 36 Big and Little Silhouette Replacements B

   > No. 37 Same-Size Silhouette Replacements

3. **Get to know your trees**

   All the tree silhouettes and leaves used in this module are representations of real trees. The tree puzzles and tree-silhouette cards have the names of the trees written on them. To learn the names of the leaves, refer to teacher master 17, *Key to Leaf Names A,* and teacher master 21, *Key to Leaf Names B.*

4. **Plan for a class tree**

   Plan ahead for the class tree. Read Investigation 1, Part 6, "A Tree Comes to Class," and teacher master 12, *Selecting and Caring for a Tree,* thoroughly. Then consult your principal and custodian for guidelines. Look to parents and the community for support. The Arbor Foundation offers good information about tree plantings (http://www.arborday.org/).

5.  **Plan for a learning center and class scrapbook**

    Students will be collecting bits and pieces of trees during the investigation of the schoolyard trees. Provide opportunities for students to observe these materials informally at a learning center (see Science Extensions at the end of Investigation 1). Two loupes/magnifying lenses are provided in the kit for students to use at the center.

    A class scrapbook with pictures of the class tree and the adopted trees and mementos of other activities can also be at the learning center. See Step 6 of Getting Ready for Investigation 1, Part 5, for instructions about how to construct a scrapbook.

6.  **Plan for science notebooks**

    See Step 8 of Getting Ready for Investigation 1, Part 1, for ways to organize the science notebooks for this module.

7.  **Plan for the word wall and pocket chart**

    As the module progresses, you will add new vocabulary words to a word wall or pocket chart and model writing and responding to focus questions. See Step 7 of Getting Ready for Investigation 1, Part 1, for suggestions about how to do this in your classroom.

8.  **Plan for focus-question charts**

    Each part of each investigation has a focus question that students are asked before and after the activity session. You'll find these questions on teacher masters 2–3, *Focus Questions A and B*. Students will glue each focus question on a page in their science notebooks and respond to it with words or drawings. At the beginning of the module, you will need to scaffold the use of notebooks. Use a chart to model how to respond to the focus question in writing or drawings. See Step 7 of Getting Ready for Investigation 1, Part 1, for suggestions on how to do this in your classroom.

9.  **Plan for letter home and home/school connections**

    Teacher master 1, *Letter to Family*, is a letter you can use to inform families about this module. The letter states the goals of the module and suggests some home experiences that can contribute to students' learning.

    There is a home/school connection for most investigations. Check the last page of each investigation for details, and plan when to make copies and send them home with students.

▶ **NOTE**
The ***Letter to Family*** and ***Home/School Connections*** are also available electronically on FOSSweb.

**TEACHING NOTE**

*For a detailed discussion of methods for working effectively with students outdoors, see the Taking FOSS Outdoors chapter in Teacher Resources.*

**TEACHING NOTE**

*Refer to Teacher Resources on FOSSweb for a list of appropriate trade books that relate to this module.*

**10. Consider safety issues indoors and outdoors**

Early-childhood students should be allowed to demonstrate that they can act responsibly with materials, but they must be given guidelines for safe and appropriate use of materials. Work with students to develop those guidelines so they can participate in making behavioral rules and understand the rationale for the rules. Emphasize that materials do not go in mouths, ears, noses, or eyes. Encourage responsible actions toward other students.

Two safety posters are included in the kit to post in the room—*Science Safety* and *Outdoor Safety*. The Getting Ready for Investigation 1, Part 1, will offer suggestions for this discussion.

Also be aware of any allergies that students in your class might have. In Investigation 4, Part 2, students explore fruit from trees and have the opportunity to taste the fruit. Be sure to use fruits that are safe for all your students.

**11. Gather books from library**

Check your local library for books related to this module. Visit FOSSweb for a list of appropriate trade books that relate to this module.

**12. Check FOSSweb for resources**

Go to FOSSweb to review the print and digital resources available for this module.

# CARE, *Reuse, and Recycling*

When you finish teaching the module, inventory the kit carefully. Note the items that were used up, lost, or broken, and immediately arrange to replace the items. Use a photocopy of the Kit Inventory List in this chapter, and put your marks in the "Equipment Condition" column. Refill packages and replacement parts are available for FOSS by calling Delta Education at 1–800–258–1302 or by using the online replacement-part catalog (www.DeltaEducation.com).

Standard refill packages of consumable items are available from Delta Education. A refill package for a module includes sufficient quantities of all consumable materials (except those provided by the teacher) to use the kit with two classes of 32 students.

Here are a few tips on storing the equipment after use.

- Make sure items are clean and dry before storing them.
- Make sure the posters and print materials are flat on the bottom of the box.
- Inventory and bag up the cards, felt leaves and shapes, and poster labels.

The items in the kit have been selected for their ease of use and durability. Make sure that items are clean and dry before putting them back in the kit. Small items should be inventoried (a good job for students under your supervision) and put into zip bags for storage. Any items that are no longer useful for science should be properly recycled.

**Part 1**
Observing Schoolyard Trees....**50**

**Part 2**
Tree Parts.............................**62**

**Part 3**
Tree Puzzles.........................**67**

**Part 4**
Tree-Silhouette Cards.............**71**

**Part 5**
Adopt Schoolyard Trees..........**75**

**Part 6**
A Tree Comes to Class...........**82**

# PURPOSE

## Content

- Trees are living plants.

- Trees have structures: branches, leaves, trunk, and roots.

- Trees differ in size and shape.

- Trees have basic needs: light, air, nutrients, water and space.

## Scientific and Engineering Practices

- Observe trees in the schoolyard, using the five senses.

- Compare trees for similarities and differences.

- Communicate observations made about different kinds of trees, orally and through drawings.

- Identify trees as resources that are used in everyday life.

- Help plant and care for a tree.

**Full Option Science System**

# INVESTIGATION **1** – *Observing Trees*

| | Investigation Summary | Time | Focus Question |
|---|---|---|---|
| **PART 1** | **Observing Schoolyard Trees**<br>Students go on a walk around the schoolyard, developing general concepts about trees and discussing how trees are useful to people and wild animals. | **Introduction**<br>20 minutes<br>**Outdoors**<br>30 minutes<br>**Notebook**<br>15 minutes | **What did we learn about our schoolyard trees?** |
| **PART 2** | **Tree Parts**<br>Students use picture and word cards to identify the main parts of trees.  Students' understanding of tree parts is enhanced as they put together their own pictures of tree parts. | **Introduction**<br>5 minutes<br>**Center**<br>15–20 minutes<br>**Notebook**<br>15 minutes | **What are the parts of trees?** |
| **PART 3** | **Tree Puzzles**<br>Students use puzzles to learn and compare the different shapes of trees. | **Introduction**<br>5 minutes<br>**Center**<br>10–15 minutes | **What shapes are trees?** |
| **PART 4** | **Tree-Silhouette Cards**<br>Students play a matching game, using matched sets of Tree-Silhouette Cards. | **Introduction**<br>5 minutes<br>**Center**<br>10–15 minutes<br>**Notebook**<br>15 minutes | **Which trees have similar shapes?** |

| Content | Writing/Reading | Assessment |
|---|---|---|
| • Trees are living plants. | **Science Notebook Entry**<br>Draw or write words to answer the focus question. | **Embedded Assessment**<br>Teacher observation |
| • Trees have structures: branches, leaves, trunk, and roots. | **Science Notebook Entry**<br>Draw or write words to answer the focus question. | **Embedded Assessment**<br>Teacher observation |
| • Trees differ in size and shape.<br>• Trees have structures: branches, leaves, trunk, and roots. | **Science Notebook Entry**<br>Complete the tree puzzles to answer the focus question. | **Embedded Assessment**<br>Teacher observation |
| • Trees differ in size and shape. | **Science Notebook Entry**<br>Organize tree silhouettes to answer the focus question. | **Embedded Assessment**<br>Teacher observation |

| | Investigation Summary | Time | Focus Question |
|---|---|---|---|
| **PART 5** | **Adopt Schoolyard Trees**<br>The class adopts several schoolyard trees to observe throughout the school year.  Students start a classroom scrapbook to document their observations. | **Introduction**<br>5 minutes<br><br>**Outdoors**<br>20 minutes<br><br>**Notebook**<br>15 minutes<br><br>**Reading**<br>15 minutes | **What can we find out about our adopted trees?** |
| **PART 6** | **A Tree Comes to Class**<br>A living tree enters the classroom.  Students learn that a tree is alive and discuss what it needs to grow and stay healthy. The whole class goes outdoors to plant the tree that they have been observing in the classroom. | **Introduction and Reading**<br>20 minutes<br><br>**Weekly Observations**<br>5–10 minutes<br><br>**Tree Planting**<br>30 minutes | **What do trees need to grow?** |

| Content | Writing/Reading | Assessment |
|---|---|---|
| • Trees differ in size and shape.<br>• Trees have structures: branches, leaves, trunk, and roots.<br>• Trees are living plants. | **Science Notebook Entry**<br>Draw or write words to answer the focus question.<br><br>**Science Resources Book**<br>"Where Do Trees Grow?" | **Embedded Assessment**<br>Teacher observation |
| • Trees are living plants.<br>• Trees have basic needs: light, air, nutrients, water, and space. | **Science Notebook Entry**<br>Draw or write words to answer the focus question.<br><br>**Posters and Story**<br>"A Tree Comes to Class" | **Embedded Assessment**<br>Teacher observation |

# BACKGROUND *for the Teacher*

**Trees** are just about everywhere. We see them in yards, in parks, on farms, by lakes and streams, and along streets on the way to school. When we look carefully, we see that each **different** kind of tree has a distinct shape to its leaves and crown, and its bark has a distinct look and feel.

## What Did We Learn about Our Schoolyard Trees?

Students go outdoors to observe schoolyard trees and are introduced to them as **living plants**. Students observe plant structures and label the parts on tree posters.

## What Are the Parts of Trees?

What defines a tree? Several features or parts are common to almost all trees. Trees always have **roots**. Roots anchor the tree to the ground and take up the water and minerals needed by the tree to make food. A mature tree may take up 200–1200 liters (L) (50–300 gallons) of water per day.

The woody main **stem** of a tree is the **trunk**. It has to be strong enough to support the crown of the tree (the leaves and branches). The trunk is covered by protective **bark**. Bark from different trees varies in thickness, hardness, **pattern**, and **texture**.

**Branches** grow from the trunk. Branches get smaller and smaller the farther they are from the trunk and main branches. Branches and small **twigs** support the **leaves**, **flowers**, and **seeds** of the tree. The branches and twigs are arranged so as many leaves as possible will receive sunlight.

Most leaves on trees are green. A chemical known as chlorophyll gives leaves their green color. The leaves with which we are most familiar are broad and flat and come in a vast variety of sizes and **shapes**. However, the needles and scales on pines, firs, redwoods, and junipers are all leaves, too.

## What Shapes Are Trees? and Which Trees Have Similar Shapes?

Most trees fall into two categories—**conifers** and **hardwoods**. The two categories can often be told apart by their gross shape, or silhouette. Many conifers have one main trunk that goes clear to the top, making them appear somewhat pointed. They have needles or scales for leaves and are usually evergreen; that is, they never lose all their leaves at the same time.

**Typical conifer shape**

Hardwood trees, also called broadleaf trees, often have a trunk that divides into several major branches, resulting in a tree that looks wider and more rounded. The leaves on hardwoods are thin, broad, and flat. In colder climates, hardwoods tend to be deciduous, meaning that they lose their leaves at the onset of winter.

## What Can We Find Out about Our Adopted Trees?

**Adopting** a tree or two has several benefits for the novice observer, in this case, the young student. Adoption implies assuming a degree of responsibility for the affairs and well-being of the trees, so students will feel a connection. As a result, they will pay more attention. It is the attention that pays the dividends. Students will **observe** when the adopted tree gets water and what happens to it when the wind blows or it snows. Students learn when the leaves turn color or fall off and when the new ones bud and leaf out. When students regularly visit a tree, they may learn that others visit the tree, too—birds, squirrels, ants, butterflies, beetles, and other students. Students may observe that flowers appear on one of their adopted trees and, later, that seeds develop and fall. The particular patterns of the bark, the shapes of the leaves and the smell of the flowers become part of students' growing knowledge of trees, shaped by the attributes (properties) of their particular adopted trees.

Typical hardwood shape

The most significant tree in the life of your students will be the young tree they get to know in the classroom and then plant in the schoolyard. This is a tree that they can grow with all year, next year, and throughout their elementary-school careers. The deep lesson is one of great importance now and into the future. The magnificent, imposing presence of a mature tree started as a rooted twig. With very little effort, a tree can be planted, nurtured through its early years, and established as an honored and respected member of a local environment.

## What Do Trees Need to Grow?

Life is a unique and precious condition of matter. Trees are alive, so trees have the same fundamental needs as all other organisms.

Trees need water. Life happens in cells, and the complex chemistry that defines life takes place in an aqueous environment. Trees also require nutrients for energy, growth, repair, and maintenance. Animals get their nutrients by consuming food, that is, other organisms. Trees, like all plants, make their own food from raw materials in the environment—minerals from the soil, gases from the air, and light and energy from the Sun. Trees produce waste products as they go about their metabolic business. They discharge their waste into the air.

In summary, trees need water and minerals from the soil, carbon dioxide and oxygen from the air, light from the Sun, and access to the air to eliminate waste. When these basic needs are met, a tree can survive and grow for years, outliving the humans that surround it by years, centuries, or millennia.

Say it
Write it
**New Word**
See it
Hear it

*Adopt*
*Bark*
*Branches*
*Circumference*
*Compare*
*Cone*
*Conifer*
*Desert*
*Different*
*Flower*
*Hardwood*
*Leaves*
*Living*
*Mountain*
*Observe*
*Ocean*
*Pattern*
*Plant*
*River*
*Roots*
*Rubbing*
*Seed*
*Shape*
*Similar*
*Stem*
*Swamp*
*Texture*
*Tree*
*Trunk*
*Twig*
*Valley*

# TEACHING CHILDREN *about*
## *Observing Trees*

The exploration of trees should be fun and exciting. You may discover that students are, in a sense, seeing trees for the first time. We need to help children at this age build a large repertoire of experiences in a very naturalistic way. The broader these experiences, the more conceptual knowledge students will have to draw on later.

The **conceptual flow** for this first investigation starts with a walk in the schoolyard to look for **trees**. Trees are identified as **plants**, and trees are confirmed as **living** plants. Students stop at one tree to observe its main **structures—roots**, **trunk**, **branches**, and **leaves**.

In Part 2, students use word and picture cards to reinforce the concepts and vocabulary associated with the major parts of trees.

In Parts 3 and 4, students work with tree puzzles and tree-silhouette cards to experience the **variety of shapes** trees assume. Students compare and sort tree silhouettes into two generalized shapes, **pointed** and **rounded**.

In Part 5, students adopt two trees and get to know more about them. They discover **additional structures** that are part of their adopted trees, including **bark**, **twigs**, **flowers**, **seeds**, and **cones**.

In Part 6, students have a small tree in class that they observe for a short time before planting it in the schoolyard. In preparation for planting, students learn that a living tree has **basic requirements for life**, including **water**, **air**, **space**, **light**, and **nutrients**.

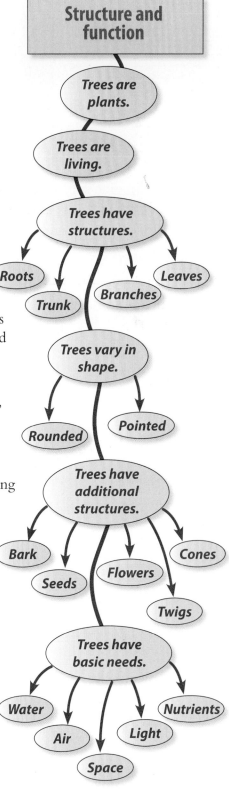

*No. 1—Teacher Master*

**FOCUS QUESTIONS A**

| | |
|---|---|
| Inv. 1, Part 1: | What did we learn about our schoolyard trees? |
| Inv. 1, Part 2: | What are the parts of trees? |
| Inv. 1, Part 3: | What shapes are trees? |
| Inv. 1, Part 4: | Which trees have similar shapes? |
| Inv. 1, Part 5: | What can we find out about our adopted trees? |
| Inv. 1, Part 6: | What do trees need to grow? |
| Inv. 2, Part 1: | What can we observe about leaves? |
| Inv. 2, Part 2: | What shapes are leaves? |
| Inv. 2, Part 3: | How are leaves different? |
| Inv. 2, Part 4: | How are leaf edges different? |
| Inv. 2, Part 5: | What can we observe about leaves? |

*No. 2—Teacher Master*

**LABELS FOR TREE POSTERS**

| TREE | TREE |
|---|---|
| branches | branches |
| leaves | leaves |
| roots | roots |
| trunk | trunk |

*No. 30—Teacher Master*

# MATERIALS *for*
## Part 1: *Observing Schoolyard Trees*

### For each student

- 1  *Letter to Family*
- 1  Science notebook (See Step 8 of Getting Ready.) ★
- •  Crayons and pencils ★

### For the class

- 1  Carrying bag ★
- 2  Tree-trunk rounds
- 1  Poster, *White Pine*, with library pocket
- 1  Poster, *Red Oak*, with library pocket
- 2  Sets of poster labels, 5/set (See Step 3 of Getting Ready.)
- 10  Round hook-and loop fasteners (See Step 3 of Getting Ready.)
- •  Glue sticks
- •  Chart paper ★
- •  Phone books or catalogs ★
- ★  Self-stick notes
- •  *Science Safety* and *Outdoor Safety* posters
- ❏  1  Teacher master 1, *Letter to Family*
- ❏  1  Teacher master 2, *Focus Questions A* (See Step 8 of Getting Ready.)
- ❏  1  Teacher master 30, *Labels for Tree Posters,* replacement masters (Optional; see Step 3 of Getting Ready.)

### For assessment

- ❏  •  *Assessment Checklist*

★ Supplied by the teacher.          ❏ Use the duplication master to make copies.

# GETTING READY *for*
## Part 1: *Observing Schoolyard Trees*

### 1. Schedule the investigation

Part 1 is a whole class-activity. Plan 15–20 minutes for the introduction when students draw a picture of a tree, 30 minutes to take the schoolyard walk and label the tree posters, and 15 minutes to record in the notebook and to wrap up.

### 2. Preview Part 1

Students go on a walk around the schoolyard, developing general concepts about trees and discussing how trees are useful to people and wild animals. The focus question is **What did we learn about our schoolyard trees?**

### 3. Check the two tree posters

Check the condition of the set of five tree-part labels (tree, branches, leaves, roots, trunk) for each poster—Red Oak and White Pine. Each label should have the loop side of a fastener dot on the back side. The poster should have the hook side of the fasteners in the five appropriate locations.

If a label is missing, copy teacher master 30, *Labels for Tree Posters*, (onto card stock if possible), laminate it, cut the label out, and stick a loop fastener on the back. The kit contains extra loop-and-hook-fasteners for this purpose.

Put a set of labels in each library pocket on the posters.

Locate a place in the classroom to put up the posters after the class has completed their walk outdoors.

**TEACHING NOTE**

*The Getting Ready section for Part 1 of Investigation 1 is longer than the corresponding section for the other parts. Several of the numbered items appear only here but may apply to other parts as well. These include planning a word wall (Step 7), photocopying duplication masters, focus questions, and center instruction sheets (Steps 9–10), and planning for safety (Step 11).*

TREE

branches

leaves

trunk

roots

4. **Plan for pressing leaves**

Gather a collection of catalogs or phone books for pressing leaves. Pressing ensures that the leaves will dry flat. If you use catalogs with shiny paper, the dry leaves remain more supple. Keep the leaves in the books until they are used in Investigation 2.

Write students' names on self-stick notes. Stick them to the edges of the pages, so they extend outside the book like index tabs.

5. **Select your outdoor site**

Get to know the trees in your schoolyard. Walk around your schoolyard and determine the route you will take with students when they look for trees. If possible, identify the kinds of trees so you will be able to name them for students. The custodian or school gardener might be a good resource for identifying the trees.

Enlarging your learning space to include the schoolyard and the local neighborhood requires implementing some new teaching methods and learning behaviors. For a more detailed discussion of these methods, see the Taking FOSS Outdoors chapter in *Teacher Resources*.

6. **Check the site**

It is always a good idea to check the outdoor site on the morning of an outdoor activity. Check for any distracting items or unsafe items where students will be working.

7. **Plan to use a word wall or pocket chart**

As the module progresses, you will add new vocabulary words (a) to cards or sentence strips for use in a pocket chart and/or (b) to a word-wall chart for posting on a wall or an easel. You will also use a chart for writing the day's focus question and for modeling responses. For additional information, see the Science-Centered Language Development chapter in *Teacher Resources*.

**TEACHING NOTE**

*For a detailed discussion of strategies for working outdoors with students, see the Taking FOSS Outdoors chapter in Teacher Resources.*

**FOCUS CHART**

*What did we learn about our schoolyard trees?*

Trees have branches and leaves.

Squirrels, birds, and other animals use trees for homes and food.

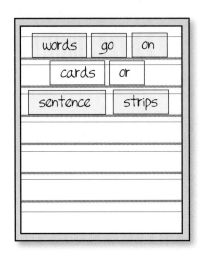

words go on

cards or

sentence strips

**WORD WALL**

words
words
and more words

### 8. Plan for student notebooks

Each student will keep a record of science investigations in his or her science notebook. Students will record observations and responses to focus questions. This record will be a useful reference document for students and a revealing testament for adults of each student's learning progress.

We recommend that students use bound composition books for their science notebooks. This ensures that student work is maintained as an organized, sequential record of student learning.

You can make individual science notebooks by stapling 12–15 blank pages together with a cover. If you are already using an alternative method of organization with your students, such as a sheaf of folded and stapled pages, your method can take the place of the bound composition book.

Each part of each investigation starts with a focus question. You'll find these questions on teacher masters 2 and 3, *Focus Questions A and B*. There are a couple of ways to provide these questions to students.

- Photocopy the teacher masters, cut the questions apart, and have them ready for each student to glue into his or her notebook at the appropriate time in each part.

- Prepare sheets with one focus question on each sheet and room for students to draw and write. You could also include writing frames on each sheet. Photocopy these sheets and assemble them into a science notebook for each student.

▶ **NOTE**
See the *Science Notebooks in K–2* chapter for details on setting up and using notebooks.

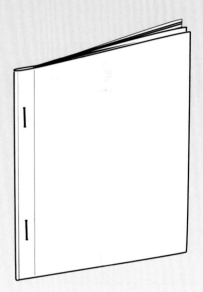

9-27-15

What did we learn about our schoolyard trees?

This tree has leaves.

4

Pine.

5

9. **Photocopy duplication masters**

    Teacher masters serve various functions—letter to family, home/school connections, center instructions, and focus questions. These can be duplicated, cut apart, and glued into student notebooks. A master that requires duplication is flagged with this icon ❑ in the materials list for each part.

10. **Send a letter home to families**

    Read teacher master 1, *Letter to Family*, and send it home with students as you begin this investigation. The letter explains what students will be doing with trees and weather and what parents can do to extend the experiences at home with their children.

    As a culminating activity in Investigation 2, students tape or glue dried leaves into their notebooks. The *Letter to Family* asks families to help students gather six to eight leaves from home to bring to school to be dried and pressed for use in Investigation 2.

11. **Plan for safety indoors and outdoors**

    Young children must be allowed to demonstrate that they can act responsibly with materials, but they must be given guidelines for safe and appropriate use of materials. Work with students to develop those guidelines so that students participate in making behavior rules and understand the rationale for the rules. Encourage responsible actions toward other students. Display and discuss the *Science Safety* and *Outdoor Safety* posters in class.

    Look for the safety icon in the Getting Ready and Guiding the Investigation sections, which alerts you to safety concerns throughout the module. Be aware of any plant allergies your students might have.

12. **Plan for working with English learners**

    At important junctures in an investigation, you'll see sidebar notes titled "EL Note." These notes suggest additional strategies for enhancing access to the science concepts for English learners. Refer to the Science-Centered Language Development chapter for resources and examples to use when working with science vocabulary, writing, oral discourse, and readings.

    Each time new science vocabulary is introduced, you'll see the new-word icon in the sidebar. This icon lets you know not only that you'll be introducing important vocabulary, but also that you might want to plan on spending more time with those students who need extra help with the vocabulary.

Say it
See it
Hear it
Write it
**New Word**

### 13. Assess progress throughout the module

Assessment opportunities are embedded throughout the module to help you look closely at students' progress. In kindergarten, those assessment opportunities involve teacher observations of students' actions with materials as well as teacher-student and student-student verbal interactions, focusing on the ideas under study. Read through the Assessment chapter for a description of the assessment opportunities.

You will find the two-page *Assessment Checklist* in the Assessment Masters chapter in *Teacher Resources*. The first page lists the science-content objectives that students will encounter throughout the module and which are described in each Getting Ready section. You can assess each of these objectives several times during the course of the module. There is also a space on the checklist for writing your observations of each student. The second page of the *Assessment Checklist* is for recording students' ability to engage in scientific practices appropriate for their age.

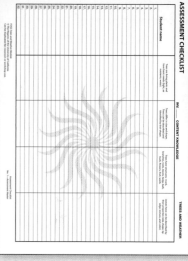

*No. 1—Assessment Master*

### 14. Plan assessment for Part 1

There are six objectives that can be assessed at any time during any part of this investigation.

**What to Look For**

- *Students ask questions.*
- *Students use their senses to observe living things.*
- *Students show respect for living things.*
- *Students record observations.*
- *Students communicate observations orally, in writing, and in drawings.*
- *Students use new vocabulary.*

Here are specific content objectives for this part.

- *Trees are living plants; trees have basic needs.*
- *Trees have structures.*
- *Trees are a natural resource.*

Focus on a few students each session. Record the date and a + or − on the *Assessment Checklist*.

Make copies of the *Assessment Checklist*, attach them to a clipboard, and carry them with you when students are engaged in the investigations. Record your observations as you interact with students, or take a few minutes after class to reflect on the lesson.

**TEACHING NOTE**

*Because there are several opportunities for you to assess students on each objective, we suggest that you focus on six to ten students during each session rather than trying to assess the whole class at one time.*

**What did we learn about our schoolyard trees?**

## GUIDING *the Investigation*
### Part 1: *Observing Schoolyard Trees*

1. **Introduce** *observe*

   Call students to the rug. If this is their first introduction to making observations of objects, introduce the word **observe** with a mini-lesson. One way to do this is to have the students respond chorally. Below is an example of this method.

   ➤ *Today we are going to learn a very important science word. Say "observe."* [Write "observe" on the word wall or on a card to place in a pocket chart.]

   S: Observe.

   ➤ *Let's clap the parts (syllables) of the word.*

   S: Ob (clap) serve (clap).

   ➤ *Again.*

   S: Ob (clap) serve (clap).

   ➤ *How many parts (syllables) are in the word observe?*

   S: Two.

   ➤ *Observe means to look at something carefully. What does observe mean?*

   S: To look at something carefully.

   ➤ *When I look at something carefully, I _____ .*

   S: Observe.

   ➤ *We can observe with our eyes, but we can also observe using our other senses. We can listen, touch, smell, and sometimes taste. Today we are going to observe something carefully by looking at it, by touching it, and by smelling it, so we are going to _____ .*

   S: Observe.

▶ **SAFETY NOTE**

Remind students never to taste anything during science class unless the teacher says it is OK to do so.

2. **Introduce the study of trees**

   Ask students several questions.

   ➤ *Who has seen a **tree** lately?*

   ➤ *Where was it?*

   ➤ *What did it look like?*

   ➤ *Where else could you go to see trees?*

   ➤ *What would you like to find out about trees?*

## 3. Distribute notebooks, pencils, and crayons

Tell students that they are going to begin by drawing a tree. Send them to tables and distribute the science notebooks, pencils, and crayons. Tell students what you want them to do.

a. *Draw a picture of a tree on page 1 of your notebooks.*

b. *Write "tree" above or below your tree drawing.*

Write "tree" on the board for students to copy and have students start their drawings. Students will repeat the tree-drawing activity in their notebooks three more times during the school year to demonstrate how their understanding of trees is progressing.

Collect the notebooks and review student drawings.

**Materials for Step 3**
- *Science notebooks*
- *Pencils and crayons*

**TEACHING NOTE**

*This first drawing will provide a record of students' pre-instructional concept of a tree. Have students draw in their notebooks before going outdoors to observe trees for the first time.*

## B R E A K P O I N T

## 4. Go outdoors

Organize your class to go outdoors. Quickly review the procedures and behavior expectations for outdoor learning using the safety posters. Discuss orderly travel to and from the schoolyard and the importance of listening to directions. Collect your tree-trunk rounds and posters with labels. Proceed to a location with a view of trees.

Help students form a sharing circle. Have them gather around you, join hands, and move backward until they are in a reasonable approximation of a circle. Then have them drop their hands and stay in their locations. Ask,

➤ *Can you see any plants? Point to them.*

Confirm or introduce the idea that trees are plants.

*Trees are **plants**. Trees are the biggest plants in the world.*

➤ *Do you see more than one tree?*

➤ *How many trees do you see? Count them.*

➤ *Are all the trees the same? How do you know?*

**Materials for Step 4**
- *Tree-trunk rounds*
- *Posters and labels*
- *Carrying bag*

## 5. Select one tree

Ask a student to select a tree. Move your sharing circle near or around the tree. Ask,

➤ *How do you know this is a tree?*

➤ *What colors do you see on this tree?*

➤ *Does this tree look young or old? Why do you think so?*

➤ *Is the tree moving?*

➤ *Is the tree making any sounds?*

Tell students,

*The big, round part of the tree that comes out of the ground is the **trunk**. The trunk is a big **stem** that supports (holds up) the rest of the tree. Most plants have a stem.*

Invite the class to crowd around the tree trunk as you ask,

➤ *What is the texture of the tree trunk? How does the tree trunk feel?*

➤ *How does it smell?*

➤ *Can you find things on the ground that may have come from the tree?*

As students find materials under the tree, identify each material. Possible finds include bark, leaves, twigs, seeds, flowers, and buds.

➤ *Is this tree like any others you can see in the schoolyard?*

➤ *Do you think trees are living? Why do you think so?*

6. **Observe schoolyard trees**
Visit other schoolyard trees. As you visit subsequent trees, ask,

➤ *Is this tree the same as our first tree?*

➤ *How is it the same? How is it **different**?*

➤ *What parts do all trees have?*

Dwell on this last question until the four main parts of a tree have been identified: trunk, **branches**, **leaves**, and **roots**. You might have to call on students' previous experience to establish the presence of underground roots, or you might be able to see some evidence of roots above the ground.

7. **Sit under a tree**
If it is appropriate, have the class sit down under one of the trees. Encourage these observations.

• Have students look up through the leaves. Ask them if they can see the Sun shining through the leaves or if the Sun is completely blocked out.

• Ask students to look at the trunk where the trunk meets the ground, then follow it up until a branch grows out from it. Have them follow the branch until it becomes a thin **twig**.

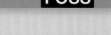

**TEACHING NOTE**

*The chance observation of an ant or spider presents an opportunity to acknowledge sometimes-maligned animals as wonderful discoveries and interesting and valued subjects for study. One goal of FOSS is to build a code of responsibility and respect for all living things.*

- Hold up the two tree rounds for students to see. Ask what students think they are. Confirm that they are slices of tree trunks called rounds and that they are from two different kinds of trees. Rounds show us what tree trunks look like inside. Give students the rounds to observe and describe.

- Ask students to look for animals such as birds, squirrels, or ants that might be on the trunk, branches, or leaves of the tree. If a spider, ant, or other insect is located, discuss the tree as a home for small animals. Remind students not to pick up any living organisms, such as bugs and insects.

8. **Use the tree posters**

   Show students the two tree posters, one of a **conifer** (white pine) tree and one of a **hardwood** tree (red oak). Ask students which poster looks more like the tree under which you are sitting.

   Call on a student to take one of the tree-part labels from the pocket on the poster. Read the label aloud, and ask him or her to stick the label onto the appropriate dot on the poster. Call on other students to repeat the process until both posters are completely labeled.

9. **Discuss how people use trees**

   Tell students several ways people use trees.

   - Trees provide shade for a cool place to rest.

   - Trees provide food.

   - Trees are cut into lumber to build homes and furniture.

   - Trees provide fuel for heat.

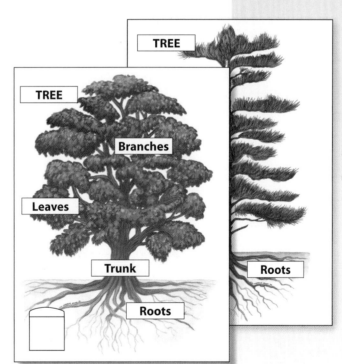

10. **Return to class**

    Return to the classroom. Put up the posters and remove the labels. Students can label the parts of the trees when they have free time.

*branch*
*conifer*
*different*
*hardwood*
*leaves*
*observe*
*plant*
*root*
*stem*
*tree*
*trunk*
*twig*

## 11. Reinforce vocabulary

Review key vocabulary used to describe trees. One way is to use a cloze review. You say a sentence, leaving the last word off, and students answer chorally. Here's an example.

➤ *When you look at something closely, you _____ .*

S: Observe.

➤ *The big plants we observed today are _____ .*

S: Trees.

➤ *The part of the tree near the ground is the _____ .*

S: Trunk.

➤ *The part of a tree that grows below the ground is the _____ .*

S: Root.

➤ *The part that grows out of the trunk is the _____ .*

S: Branch.

➤ *The branches get thinner at the end and become _____ .*

S: Twigs.

➤ *The parts that grow out from the twigs are the _____ .*

S: Leaves.

## 12. Focus question: What did we learn about our schoolyard trees?

Write the focus question on the chart as you read it aloud.

➤ *What did we learn about our schoolyard trees?*

Tell students that you have a strip of paper with the question written on it. Describe and model how to glue the strip into the notebook. Have students use pictures and words to answer the question.

## 13. Model responses to the focus question (optional)

Depending on students' experiences with notebooks, you can let them work on their own, or you can model making a notebook entry, using the focus chart.

---

**FOCUS CHART**

*What did we learn about our schoolyard trees?*

Trees have branches and leaves.

Squirrels, birds, and other animals use trees for homes and food.

---

# WRAP-UP/WARM-UP

### 14. Share notebook entries

Conclude Part 1 or start Part 2 by having students share notebook entries. Ask students to open their science notebooks to the first entry. Read the focus question together.

➤ *What did we learn about our schoolyard trees?*

Ask students to pair up with a partner to

- share their answers to the focus question;
- explain their drawings.

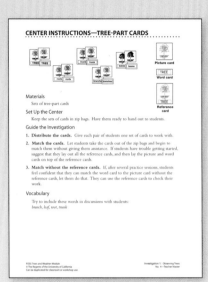

*No. 4—Teacher Master*

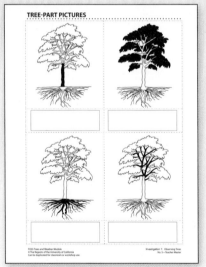

*No. 5—Teacher Master*

# MATERIALS *for*

## Part 2: *Tree Parts*

### For each pair of students at the center

- 1 Set of Tree-Part Cards, 15 cards/set
- ❏ 2 Teacher master 5, *Tree-Part Pictures*

### For the class

- 8 Zip bags, 1 L (1 qt.)
- • Glue sticks
- • Crayons and pencils
- ❏ 1 Teacher master 4, *Center Instructions—Tree-Part Cards*
- ❏ • Teacher masters 31–32, *Tree-Part Card Masters A* and *B* (Optional; see Step 3 of Getting Ready.)

### For assessment

- • *Assessment Checklist*

❏ Use the duplication master to make copies.

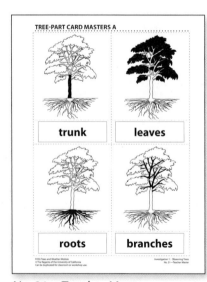

*No. 31—Teacher Master*

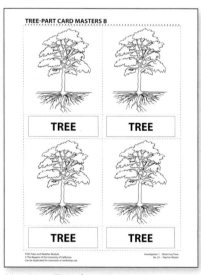

*No. 32—Teacher Master*

# GETTING READY *for*

## Part 2: *Tree Parts*

### 1. Schedule the investigation

Work with groups of six to eight students (three to four pairs) at a center. Plan a 5-minute whole-class introduction, 15–20 minutes for each group, and 15 minutes for working in notebooks as a class.

### 2. Preview Part 2

Students use picture and word cards to identify the main parts of trees. Students' understanding of tree parts is enhanced as they put together their own pictures of tree parts. The focus question is **What are the parts of trees?**

### 3. Check the tree-part cards

The kit contains eight sets of laminated Tree-Part Cards. One set (15 cards total) is illustrated below. Each set of 15 cards should be in its own zip bag.

Check that each set of cards is complete—one picture card, one word card, and one reference card with a picture and word on it for each of the five categories. Use teacher masters 31–32, *Tree-Part Card Masters*, to make copies to replace any missing cards.

> **TEACHING NOTE**
>
> *In Parts 2–5, students work at centers with representational materials. Keep the word wall and focus chart near the center, so you can add words and sentences that students generate.*

> **TEACHING NOTE**
>
> *The solid black part on each picture card identifies the plant structure on the matching word card. The TREE card is the only picture card with no solid black part.*

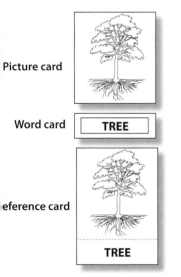

Picture card

Word card    TREE

Reference card    TREE

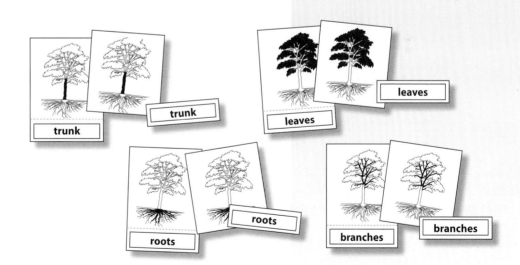

trunk

trunk

leaves

leaves

roots

roots

branches

branches

4. **Plan assessment**

Here is the specific objective to observe in this part.

- *Trees have structures.*

Focus on a few students each session. Record the date and a + or – on the *Assessment Checklist*.

# GUIDING *the Investigation*

## Part 2: *Tree Parts*

1. **Review the parts of the tree**

   Have the class look at the posters and review the parts of trees—trunk (stem), branches, roots, and leaves. Tell students that the new activity at the center will give them practice naming the parts of trees and looking at the words that represent those parts.

2. **Use the tree-part cards at a center**

   Send six to eight students to the center. They can work by themselves or with a partner. Make sure they have enough room to spread out the 15 Tree-Part Cards in each set.

   Give each student (or pair of students) a set of 15 Tree-Part Cards. Using a picture card, they should look at the part of the tree that is solid black and then find the word card for that tree part. If students have trouble getting started, suggest that they lay out all the reference cards, and then lay the matching picture and word cards on top of each reference card.

3. **Match without the reference cards**

   If, after several practice sessions, students feel confident that they can match the word cards to the picture cards without using the reference cards, let them do that. They can use the reference cards to check their work. This ends the students' work at the center.

4. **Focus question: What are the parts of trees?**

   With the whole class write or project the focus question on the chart as you read it aloud.

   ➤ *What are the parts of trees?*

   Tell students that you have a strip of paper with the focus question written on it. Describe how to glue the strip into the notebook.

   Give each student a copy of *Tree-Part Pictures* (teacher master 5). Have them glue the pictures into their notebooks

   - as a whole sheet, folded in half, or
   - as individual pictures, one or two pictures to a page.

   Have students write the word that refers to the part of the tree that is emphasized in black. They can color the trees if they wish.

## FOCUS QUESTION

*What are the parts of trees?*

**Materials for Step 2**
- *Tree-Part Card sets*

**Materials for Step 4**
- ***Tree-Part Pictures*** *sheets*
- *Glue sticks*
- *Crayons and pencils*

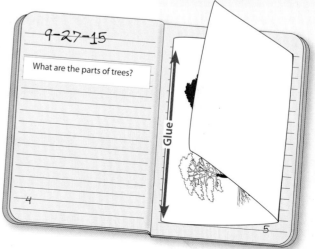

## WRAP-UP/WARM-UP

**FOCUS CHART**

*What are the parts of trees?*

branches

leaves

roots

trunk

5. **Share notebook entries**

Conclude Part 2 or start Part 3 by having students share notebook entries. Ask students to open their science notebooks to the most recent entry. Read the focus question together.

➤ *What are the parts of trees?*

Ask students to pair up with a partner to share their tree pictures and labels.

# MATERIALS *for*
## Part 3: *Tree Puzzles*

### For each group at the center

- 8 Tree puzzles (apple, cottonwood, fir, maple, oak, palm, pine, poplar), 6 pieces/puzzle, with reference sheets
- 2 Tree puzzles (cottonwood, pine), 9 pieces/puzzle, with reference sheets
- 10 Puzzle frames, clear plastic

### For the class

- 1 Poster with labels, *White Pine*
- 1 Poster with labels, *Red Oak*
- ❏ 1 Teacher master 6, *Center Instructions—Tree Puzzles*

### For assessment

- • *Assessment Checklist*

❏ Use the duplication master to make copies.

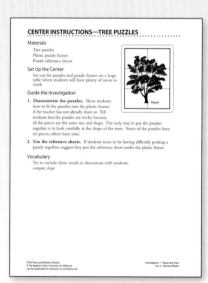

*No. 6—Teacher Master*

## GETTING READY *for*
### Part 3: *Tree Puzzles*

1.  **Schedule the investigation**
    Work with groups of six to ten students at a center. Plan a 5-minute whole-class introduction followed by 10–15 minutes for each group at the center.

2.  **Preview Part 3**
    Students use puzzles to learn and compare the different shapes of trees. The focus question is **What shapes are trees?**

3.  **Plan to demonstrate the puzzles and frames**
    The tree puzzles are tricky because all the pieces are the same size and shape. Students must pay attention to the shapes of the trees in order to put the puzzles together. Practice putting one of the puzzles together. Mix the pieces up, then put them together again by fitting them into the plastic frame.

    Eight of the puzzles have six pieces. The trees are
    *   Apple
    *   Cottonwood
    *   Fir
    *   Maple
    *   Oak
    *   Palm
    *   Pine
    *   Poplar

    Two of the puzzles have nine pieces. The trees are
    *   Cottonwood
    *   Pine

    Each puzzle has a reference sheet that can be put under the frame for students who may need assistance.

4.  **Plan assessment**
    Here are specific objectives to observe in this part.

    *   *Trees have structures.*

    *   *Students compare structures of trees.*

    Focus on a few students each session. Record the date and a + or – on the *Assessment Checklist*.

# GUIDING *the Investigation*
## Part 3: *Tree Puzzles*

**FOCUS QUESTION**
*What shapes are trees?*

1. ### Introduce *compare*
   Call students to the rug. Introduce **compare** with a mini-lesson. Below is an example of choral response.

   ➤ *Today we are going to learn a very important science word. Say "compare."* [Write the word on the word wall.]

   S: Compare.

   ➤ *Let's clap the parts (syllables) of the word.*

   S: Com (clap) pare (clap).

   ➤ *Again.*

   S: Com (clap) pare (clap).

   ➤ *How many parts (syllables) are in the word compare?*

   S: Two.

   ➤ *Compare means to look at two different things and observe how they are the same and different. What does compare mean?*

   S: To look at two different things and observe how they are the same and different.

   ➤ *When I look at two different things and tell how they are the same and different, I _____ .*

   S: Compare.

2. ### Review the posters
   Have students compare the posters of the white pine and red oak. Ask them to focus on the general **shape** of the trees. Tell them,

   *One property of a tree is its shape. I've noticed that the shape of the pine tree is different from the oak tree. People who study trees become so familiar with the shapes of trees that they can identify them (tell them apart) from far away, by looking at their shapes.*

3. ### Introduce the center
   Show students one set of puzzle pieces and a clear plastic puzzle frame. Demonstrate how to use the frame by putting in a few of the puzzle pieces.

   Tell students,

   *These puzzles are tricky because all the pieces are the same size and shape. The only way to put the puzzle together is to look carefully at the shape of the tree. Some of the puzzles have six pieces, and some have nine pieces.*

**Materials for Steps 2–3**
- *Tree posters*
- *Puzzles with reference sheets*
- *Frames*

Show students a reference sheet and tell them that if they get stuck, they can put the reference sheet under the frame to help them put the puzzle together.

4. **Send students to the center**

   Have students rotate through the center, giving each group 10–15 minutes to work with the puzzles.

   Have the tree puzzles available for students throughout their study of trees.

5. **Focus question: What shapes are trees?**

   After students have had a chance to work with the puzzles, gather them at the rug. Tell them,

   *The puzzles show different trees.*

   Introduce the focus question and write it on the focus chart as you read it aloud.

   ➤ *What shapes are trees?*

6. **Answer focus question**

   Hold up two puzzle reference sheets and ask students to compare the shapes. Add these words to the focus chart. Students can also glue a focus-question strip into their notebooks and answer it with words or drawings.

# WRAP-UP/WARM-UP

7. **Share notebook entries**

   If students make a notebook entry, you can conclude Part 3 or start Part 4 by having students share notebook entries. Ask students to open their science notebooks to the most recent entry. Read the focus question together.

   ➤ *What shapes are trees?*

   Ask students to pair up with a partner to share their drawings and words.

---

**FOCUS CHART**

*What shapes are trees?*

round and short

thin and tall

---

# MATERIALS *for*

## Part 4: *Tree-Silhouette Cards*

### For each pair of students

    2  Sets of *Tree-Silhouette Cards*, 8 cards/set

### For the class

❏   1  Teacher master 7, *Center Instructions—Tree-Silhouette Cards*

❏   1  Teacher master 8, *Tree-Silhouette Cards* (See Step 3 of Getting Ready.)

### For assessment

• *Assessment Checklist*

❏ Use the duplication master to make copies.

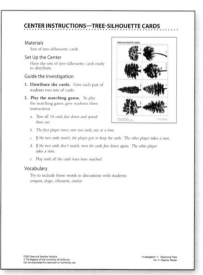

*No. 7—Teacher Master*

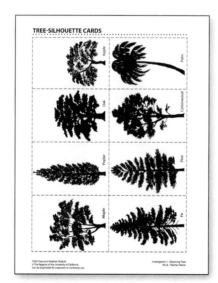

*No. 8—Teacher Master*

## GETTING READY *for*

### Part 4: *Tree-Silhouette Cards*

1. **Schedule the investigation**

   You can work with the whole class at one time or work with groups of six to ten students at a center. Plan a 5-minute whole-class introduction followed by 10–15 minutes for each group at the center. Bring the class together for 15 minutes of notebook work.

2. **Preview Part 4**

   Students play a matching game, using matched sets of Tree-Silhouette Cards. The focus question is **Which trees have similar shapes?**

3. **Prepare the Tree-Silhouette Cards**

   Use teacher master 8, *Tree-Silhouette Cards*, to make one set of eight cards for each student. At the end of the session, students can glue the cards into their notebooks.

4. **Plan assessment**

   Here are specific objectives to observe in this part.

   - *Students compare structures of trees.*

   - *Students compare shapes of trees.*

   Focus on a few students each session. Record the date and a + or – on the *Assessment Checklist*.

# GUIDING *the Investigation*
## Part 4: *Tree-Silhouette Cards*

1. **Discuss tree shapes**
   Call students to the rug. Tell them,

   *Sometimes you can identify a tree just by looking at the shape of the whole tree. Shape is a property of a tree.*

   Show students one set of Tree-Silhouette Cards. Discuss the shapes of the different trees. (Students should be familiar with most of the trees, since they are the same trees that are on the tree puzzles.) Each pair will work with two sets of Tree-Silhouette Cards.

2. **Demonstrate the activity**
   Tell students that they will work with a partner. Each pair will have two sets of the Tree-Silhouette Cards. They will play a memory matching game with the cards.

   Spread out two set of cards on the floor. Show and describe how to get started.

   a. *Turn all 16 cards face down, and spread them out.*

   b. *The first player turns over any two cards, one at a time.*

   c. *If the cards match, the player gets to keep the cards. The other player takes a turn.*

   d. *If the cards don't match, turn the cards face down again. The other player takes a turn.*

   e. *The two players take turns until all the cards have been matched.*

3. **Begin the activity**
   Send students to tables and distribute two sets of *Tree-Silhouette Cards* to each pair of students. Let the action begin. Students will need 10–15 minutes to play a couple of rounds of the game.

4. **Focus question: Which trees have similar shapes?**
   Write the focus question on the chart as you read it aloud.

   ➤ *Which trees have similar shapes?*

   Distribute the strips of paper with the focus question for students to glue into their notebooks.

---

**FOCUS QUESTION**

*Which trees have similar shapes?*

**EL NOTE**

*Review basic shapes before introducing the cards.*

**Materials for Step 3**
- *Sets of Tree-Silhouette Cards*

**FOCUS CHART**

*Which trees have similar shapes?*

*Some trees have a point on top.*

*Some trees are rounded on top.*

New Word
Say it · See it · Hear it · Write it

**TEACHING NOTE**

*For students who need help getting started, point to one silhouette, such as the pine, and ask them to find other trees that have a similar shape. Repeat the prompt as necessary.*

5. **Sort tree silhouettes**

Tell students,

**Similar** *is a word that means having the same property. One way that chairs are similar is that they all have four legs. Some shoes are a similar color, such as red. Some trees have similar shapes.*

*Use your set of eight Tree-Silhouette Cards. Sort them into two piles. The trees in each pile should be similar. When the sorting is done, glue the silhouettes into your notebook.*

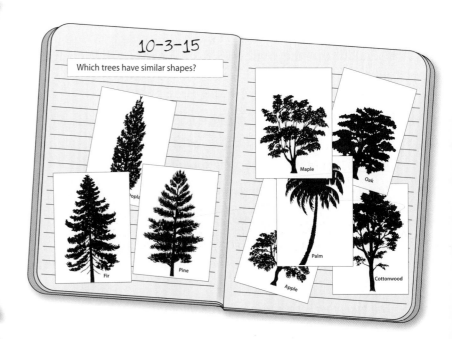

**TEACHING NOTE**

*See the **Home/School Connection** for Investigation 1 at the end of the Interdisciplinary Extensions section. This is a good time to send it home with students.*

# WRAP-UP/WARM-UP

6. **Share notebook entries**

Conclude Part 4 or start Part 5 by having students share notebook entries. Ask students to open their science notebooks to the most recent entry. Read the focus question together.

➤ *Which trees have similar shapes?*

Ask students to pair up with a partner to

- share their answers to the focus question;
- explain how they grouped the tree pictures.

# MATERIALS *for*
## Part 5: *Adopt Schoolyard Trees*

### For each student

- 1  *FOSS Science Resources: Trees and Weather*
  - "Where Do Trees Grow?"

### For the class

- 1  Camera ★
- •  String
- •  Crayons ★
- •  White paper ★
- 2   Zip bags, 1 L
- 1  Scissors ★
- 1  Permanent marking pen, black ★
- 1  Clipboard ★
- 1  Scrapbook (See Step 6 of Getting Ready.) ★
  - 20 Pieces of construction paper, different colors, 30 × 45 cm (12" × 18") ★
  - 2  Pieces of cardboard, 31 × 46 cm (12.25" × 18.25")  ★
  - •  Yarn, cord, or ribbon ★
  - •  Contact paper ★
  - 1  Hole punch
- •  White glue ★
- •  Transparent tape ★
- 2  Sets of Landforms Cards, 24 cards/set
- 1  Big book, *FOSS Science Resources: Trees and Weather*
- ❏  1  Teacher master 9, *Tree Observations*

### For assessment

- •  *Assessment Checklist*

★ Supplied by the teacher.          ❏ Use the duplication master to make copies.

**TREE OBSERVATIONS**

*No. 9—Teacher Master*

## GETTING READY *for*
### Part 5: *Adopt Schoolyard Trees*

1. **Schedule the investigation**

   This is a whole-class activity. Plan 5 minutes for the introduction, 15-20 minutes for the schoolyard walk to observe trees and collect samples, and 15 minutes for the class discussion and notebook writing. In addition, plan 15 minutes for the reading.

2. **Preview Part 5**

   The class adopts several schoolyard trees to observe throughout the school year. Students start a classroom scrapbook to document their observations. The focus question is **What can we find out about our adopted trees?**

3. **Record student observations**

   Make copies of teacher master 9, *Tree Observations*, and record an observation from each student while you are outdoors visiting the adopted trees. Use these observations to make a class book (see the Interdisciplinary Extensions section) as well as to assess student learning.

4. **Select your outdoor site**

   The class will adopt two (or more) trees. Preview the trees in or near the schoolyard, and make a list of possible trees. Good candidates for adoption have these attributes.

   - Different kinds of trees so comparisons over the year are possible—hardwood versus conifer, evergreen versus deciduous.

   - Accessible to students during the day.

   - Standing alone so that the shape of the tree can be observed.

   The location of these trees will determine the path of the tree walk with students.

   Determine how to involve students in the adoption process.

   Plan to carry a bag of equipment with you on the tree walk.
   - A camera.

   - String and scissors for measuring tree circumferences.

   - Crayons, for making rubbings.

   - *Tree Observations* sheets to record students' observations and questions.

5. **Check the site**

   It is always a good idea to check the outdoor site on the morning of an outdoor activity. Check for any distracting items or unsafe items where the students will be working.

6. **Plan a class scrapbook**

   Keep a scrapbook of observations and artifacts throughout the investigation of trees—photographs, leaves, bark, twigs, drawings, and so on. Set up three or four pages for each tree, and add pictures and mementos throughout the year. Or plan a page for each seasonal outing, keeping pictures of the adopted trees on the same page and comparing their changes throughout the seasons.

   You can make a scrapbook using 20 large pieces of construction paper (30 × 45 cm)—a different color for each page or section. Cover two pieces of cardboard with contact paper for the front and back covers. Use scissors to score the hinge side of the front cover so the book can be easily opened. Punch two holes in each piece of cardboard and paper, and tie the whole book together with colored yarn, cord, or ribbon.

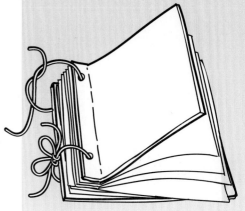

7. **Plan to read** *Science Resources*: **"Where Do Trees Grow?"**

   Plan to read "Where Do Trees Grow?" during a reading period after completing this part.

   Have one set of Landform Cards (24 cards) ready to use at the end of the reading. The images on the cards represent six kinds of landforms—mountain, valley, river, ocean, desert, and swamp. You will distribute one card to each student.

8. **Plan assessment**

   Here are specific objectives to observe in this part.

   - *Trees have structures.*

   - *Trees are living things; trees have basic needs.*

   - *Trees have different shapes.*

   Focus on a few students each session. Record the date and a + or − on the *Assessment Checklist*.

**FOCUS QUESTION**

**What can we find out about our adopted trees?**

Say it
New Word
See it
Hear it
Write it

**Materials for Step 3**

- *Clipboard*
- **Tree Observations** *sheets*
- *Camera*
- *String*
- *Scissors*
- *Crayons*
- *Paper*
- *Permanent marking pen*
- *Zip bags*
- *Carrying bag*

**TEACHING NOTE**

*Encourage students to gather samples from the ground rather than taking them from the tree.*

## GUIDING *the Investigation*
### Part 5: *Adopt Schoolyard Trees*

1. **Review schoolyard trees**
   Ask students to think back to when they looked at schoolyard trees at the beginning of the tree study. Ask,

   ➤ *How many trees do we have in the schoolyard?*

   ➤ *Are they all the same kind? How do you know?*

2. **Prepare to go outdoors**
   Tell students,

   *Today we are going on a walk outdoors to observe trees around the schoolyard again. On this walk, we will look for two different kinds of trees to* **adopt**.

3. **Go outdoors**
   Review the rules for going outdoors. Take the class on a walk around the schoolyard. Discuss the trees that students see. As they look at the trees, ask students questions about the schoolyard trees.

   ➤ *What can you tell me about this tree?*

   ➤ *Do you think this is an old tree or a young tree? How can you tell?*

   ➤ *Do you notice anything unusual about this tree?*

   ➤ *Is the tree a* **living** *or nonliving thing?*

   ➤ *Would you like to adopt this tree?*

4. **Adopt a tree**
   When you arrive at one of the trees you and your students have chosen to adopt, observe, collect, and record some data.

   - Take pictures of the tree with students gathered around. Take pictures of the leaves, **bark**, and other tree features.

   - Try to reach around the trunk of the tree. Find out how many students must hold hands to reach around the tree.

   - Mark a small spot on the tree trunk with a permanent marking pen. Wrap a string around the trunk of the tree at the level of the mark. Cut the string to record the **circumference** of the tree. (The permanent mark will allow you to remeasure the tree trunk later in the year. The mark will not harm the tree.)

   - Make crayon **rubbings** of bark **textures**. Have two students hold a piece of paper firmly against a tree while a third student rubs the side of a peeled crayon on the paper.

   - Gather samples (leaves, **flowers, seeds**, bark, twigs, **cones**, etc.) from around the tree for the scrapbook.

**5. Review vocabulary**

After you return to the classroom, have students help you add the materials that were gathered to a tree scrapbook. Include the photographs of each tree. Tape one end of the circumference string to a picture of the tree. Keep the scrapbook where students can review it freely.

This is a good time to review and add vocabulary to the word wall.

**6. Focus question: What can we find out about our adopted trees?**

Write the focus question on the chart as you read it aloud.

➤ *What can we find out about our adopted trees?*

Work with the class to generate a list of questions that students could study during the year. As students generate ideas, add them to the focus chart.

- What kind of tree is it? How old is it?
- Was the tree planted after the school was built or was the school built around the tree?
- What lives in the tree? What lives on the tree?
- Do the leaves fall off or stay on all the time?
- What happens to the tree during each season?

After a list of questions has been generated, read the focus question at the top of the chart. Tell students that you have a strip of paper with the focus question written on it. Distribute the strips for students to glue into their notebooks.

Ask students to write about or draw one thing they would like to find out about one of their adopted trees. They could also dictate it to an adult. Some students may just want to draw the tree.

# WRAP-UP/WARM-UP

**7. Share notebook entries**

Conclude Part 5 or start Part 6 by having students share notebook entries. Ask students to open their science notebooks to the most recent entry. Read the focus question together.

➤ *What can we find out about our adopted trees?*

Ask students to pair up with a partner to

- share their answers to the focus question;
- explain their drawings.

adopt
bark
circumference
cone
flower
living
rubbing
seed
texture

**Materials for Step 5**
- *White glue*
- *Transparent tape*

**FOCUS CHART**

*What can we find out about our adopted trees?*

Do the leaves fall off in winter?

What happens to the tree in a storm?

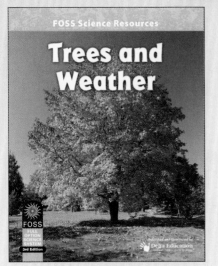

**FOSS Science Resources**

## Trees and Weather

FOSS
FULL OPTION SCIENCE SYSTEM
3rd Edition

Published and distributed by
Delta Education

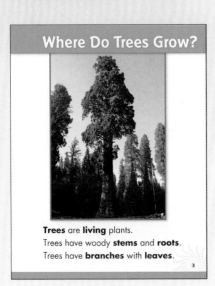

### Where Do Trees Grow?

**Trees** are **living** plants.
Trees have woody **stems** and **roots**.
Trees have **branches** with **leaves**.

3

Say it

**New Word**

Hear it

See it

Write it

## READING *in Science Resources*

8. **Preview the book cover**

Give students a few minutes to look at and discuss the cover of the book. Ask students to describe the tree in the photo.

Introduce the table of contents and read the article titles. Tell students that this book may provide answers to some of their questions about their adopted trees.

9. **Read "Where Do Trees Grow?"**

Students have observed trees in their immediate environment and compared tree shapes. This first article expands students' knowledge by introducing a variety of environments where trees live. Students review the characteristics of different landforms and learn that some trees survive naturally, while others are cared for by people.

Ask students where trees grow. Brainstorm a few answers, encouraging students to think beyond their immediate surroundings. Ask if they have seen trees when they went to visit another place. Where were the trees growing? Do trees grow in the mountains? Do they grow near the ocean? Make a list of responses on chart paper.

Introduce the title, "Where Do Trees Grow?" Explain that this article will tell students about the different places that trees grow. Ask students to listen for ideas to add to the brainstorming list.

Read the article aloud. Pause to discuss key points.

*Page 4*    *Have you seen a tree in the city?*

*Page 5*    *Have you seen a tree in the woods or a forest?*

*Page 6*    *Describe what a **mountain** looks like.*

*Page 7*    *Describe what a **valley** looks like. How is it different from a mountain?*

*Page 8*    *Describe what a **river** is like.*

*Page 9*    *How is the **ocean** different from a river?*

*Page 10*   *Describe what a **desert** is like.*

*Page 11*   *How is a **swamp** different from a desert?*

*Page 12*   *Have you seen a tree in an orchard?*

*Page 13*   *Do we have trees that are planted and cared for by people at our school?*

## 10. Discuss the reading

Discuss the reading, using these questions as a guide.

➤ *Do we have mountains or valleys where we live?*

➤ *Do we have deserts or swamps?*

➤ *Is there a river nearby?*

➤ *Is the ocean nearby?*

## 11. Distribute landforms cards

Shuffle the 24 cards in one set of Landforms Cards and enough cards from the second set to distribute one card to each student. Each student should study their own card and then compare it with the card of the person sitting next to him or her. Pairs of students should describe how the landforms are the same and how they are different.

## 12. Read "Where Do Trees Grow?" a second time

Tell students that you are going to reread the article. After you read about a landform, you will ask students with that landform to hold up their cards.

Stop reading at the end of page 6. Ask students who have cards showing mountains to hold up their cards for all to see. Continue, sharing cards after reading about each landform.

## 13. Sort landforms cards

Set up a center where pairs of students can sort the cards based on the type of landform.

*No. 10—Teacher Master*

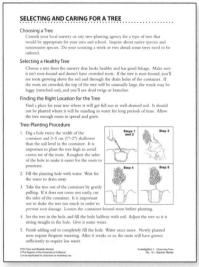

*No. 11—Teacher Master*

## MATERIALS *for*

### Part 6: *A Tree Comes to Class*

#### For the class

- 1  Tree (See Step 4 of Getting Ready.) ★
- •  Poster set, *A Tree Comes to Class* (4 posters, optional)
- 1  Camera ★
- •  String
- •  Scissors ★
- •  Transparent tape ★
- •  White glue ★
- •  Water ★
- 1  Tree scrapbook (from Part 5)
- •  Tools for planting (shovel, hose, bucket) ★
- 1  Teacher master 10, *A Tree Comes to Class—Page 1*
- 1  Teacher master 11, *A Tree Comes to Class—Page 2*
- 1  Teacher master 12, *Selecting and Caring for a Tree*

#### For assessment

- •  *Assessment Checklist*

★ Supplied by the teacher.

*No. 12—Teacher Master*

# GETTING READY *for*

## Part 6: *A Tree Comes to Class*

### 1. Schedule the investigation

Plan 20–30 minutes for the reading and the introduction of the new tree. In addition, plan a few minutes of observation several days each week for 2 weeks, and 30 minutes for a planting ceremony in the schoolyard.

### 2. Preview Part 6

A living tree enters the classroom. Students learn that a tree is alive and discuss what it needs to grow and stay healthy. The whole class goes outdoors to plant the tree that they have been observing in the classroom. The focus question is **What do trees need to grow?**

### 3. Think about planting a tree at your school

In this part, students observe a living tree for 2 weeks in the classroom and then plant it on the school grounds. Talk with your principal and the custodial or grounds manager for approval, and find out if there are restrictions on where the tree can be planted.

If, for some reason, you are not allowed to plant a tree at the school, think about alternatives. You might plant the tree in a park or near the public library where the entire community can enjoy it. Call upon your local city and county resources to help you decide what kind of tree to plant. Alternatively, you might have a tree *visit* your classroom for a few weeks. Nurseries may loan you a tree for the 2-week observation period, or you could bring in a small tree that does well indoors. Fig trees (*Ficus benjamina*) and Norfolk Island pines (*Araucaria excelsa*) do well indoors.

### 4. Obtain a tree

Container trees are available at most nurseries. Try to find a tree that is 1.2–2.5 meters (4–8 feet) tall. It may be possible to get a tree donated if you contact a nursery and explain your science project. Or ask a nursery worker if he or she has a tree that can be obtained at a reduced price. If you cannot plant a tree at your site, ask the nursery to lend you a tree for a 2-week visit.

Consider fruit trees, evergreens, colorful autumn trees, and flowering spring trees. See teacher master 12, *Selecting and Caring for a Tree*, for suggestions.

> **TEACHING NOTE**
>
> *The Arbor Day Foundation is a nonprofit conservation and education organization. A million members, donors, and partners support their programs to make our world greener and healthier. For information on community projects involving tree planting, check the website (www.arborday.org).*

5. **Identify a place for the tree in the classroom**

   Plan where you will keep your tree in the classroom or in an outdoor area near your classroom for 1–2 weeks. It should be readily accessible for continued observations. The tree will do fine in the classroom for 1–2 weeks without direct sunlight. You may want to prop it up on wood blocks if the floor is heated at your school.

6. **Plan to add to scrapbook**

   Plan to add information about the new tree to the class scrapbook. Include photographs, leaves, bark rubbings, twigs, drawings, and so on.

7. **Prepare for planting**

   Choose where you will plant your tree. Use teacher master 12, *Selecting and Caring for a Tree,* to guide the procedure. For more detailed instructions on the particular kind of tree you are planting, consult a local nursery.

8. **Plan a ceremony**

   Make planting the tree a momentous occasion. A day or two before the big event, have students plan a ceremony. Be sure to invite the principal and plan a tree dedication.

9. **Plan to read "A Tree Comes to Class"**

   Introduce this part with the story "A Tree Comes to Class." The text for the story can be found on teacher masters 10 and 11, *A Tree Comes to Class—Part 1* and *Part 2*. A set of color posters is included to provide visuals for the story as you read.

10. **Plan assessment**

    Here is a specific objective to observe in this part.

    - *Trees are living plants; trees have basic needs.*

    Focus on a few students each session. Record the date and a + or – on the *Assessment Checklist.*

# GUIDING *the Investigation*
## Part 6: *A Tree Comes to Class*

1. **Read a story to the class**

   Call students to the rug. Read the story "A Tree Comes to Class." Choose names for the children in the story that are not the names of students in your class.

   As the story progresses, introduce each of the four posters. Pause and ask questions to involve students in the story by having them recall some of the experiences they had when they went on their schoolyard walks.

   *It was Tuesday morning. Michael bounded down the stairs and hurried into the kitchen. He had put on his best shirt and his new sneakers. He was more excited today than he had been in a long time.*

   *"Whoa," said his mother. "What is so special today? You're all dressed up and in a big hurry. Usually, I have to call you at least three times before you'll get ready for school."*

   Introduce Poster 1 here.

   *Michael sat in his chair at the table and put his favorite cereal in his bowl. "Today is a special day, Mom. Mr. Garcia said we are going to get a baby tree in our class, and we are going to plant it at our school in a couple of weeks."*

   *"I'm glad your class gets to plant a tree," said Michael's mom. "What kind of tree are you going to plant?"*

   *"I don't know. Maybe it will be a . . . um . . ." Michael looked out the window and saw a big pine tree in the backyard. "Maybe it will be a pine tree . . . , but I don't know for sure."*

   *Michael ate his breakfast faster than usual. He didn't want to be late for the bus.*

   *Michael put the sandwich his mom had prepared into his lunchbox. He added some carrot sticks and selected a beautiful red apple from the basket on the table.*

   *"You know," said Michael's dad, "that apple grew on a tree."*

   *Michael thought, "Hmm, apples grow on trees. I wonder if Mr. Garcia will bring us an apple tree."*

**FOCUS QUESTION**

*What do trees need to grow?*

*Poster 1*

**EL NOTE**

*Point to objects in the pictures as you read (e.g., pine tree, apple, etc.).*

Poster 2

Poster 3

*Michael grabbed his lunch and headed out the door with his dad. He could see two trees by the street where he waited for the bus. He turned to his dad and asked, "What kind of trees are those by the edge of our street?"*

Introduce Poster 2 here.

*"I'm not really sure," said his dad, "but I do know that they give lots of nice shade in the summer."*

*Michael wondered if Mr. Garcia would bring a young shade tree to plant at school.*

*Just then, the school bus rolled around the corner. Michael ran to the bus stop and climbed on board. On the way to school, he saw trees everywhere. He never noticed before how many different kinds of trees there were.*

Pause to ask students to name as many kinds of trees as possible. Tell them that the trees they mention might be coming to Michael's class but that you will have to read on to find out.

*Today was a very special day for Michael's class. Each year, Mr. Garcia's class got a young tree to plant on the school grounds. This had been happening for 5 years, and along one side of the school grounds there now stood five trees, each one a different size.*

*When all the children had arrived and were seated on the rug, Mr. Garcia asked, "Who remembers why this is a special day?"*

*Everybody's hand went up as high as they could reach. Mr. Garcia called on Shawna. She said, "Because we get a tree today!"*

*Cory added, "And we get to have it right here in our room."*

*"That's right," said Mr. Garcia. He walked over to the corner and brought out a big can with a young tree in it.*

*"Oooooo," exclaimed all the children at the same time.*

Introduce Poster 3 here.

*When the students saw their tree, they had many questions. Michael wanted to know what kind of tree it was. Beth asked how the tree got there. Shawna wanted to know if it really was a tree because the trunk was so skinny. Cory wondered if the tree was a living thing and what it would need to grow.*

*Mr. Garcia told the students, "This tree is a mulberry tree. It came from the Sunshine Tree Farm on the other side of town."*

*"A mulberry tree!" thought Michael. "I wonder what a mulberry tree looks like when it gets big." All the students were happy to have a mulberry tree as a new member of the class.*

Pause to ask students if they can think of other questions the students might have asked Mr. Garcia about the tree. Make a list of the questions to use to study your class tree in the weeks ahead.

*At lunchtime, Mr. Garcia asked Michael if he would like to take the tree outside for some sun. Mr. Garcia helped Michael lift the little tree into the wagon. Mrs. Yee, the principal, saw Michael and said, "I see you've got a mulberry tree, and you're going to be planting it in the schoolyard soon."*

*"That's right!" said Michael. "I wonder where we should plant it. What about over there by those big shade trees?"*

Introduce Poster 4 here.

*"That would be a good place," said Mrs. Yee. "A hose would reach far enough so you'd be able to water it. And those big mulberry trees would keep it company."*

*"Those are mulberry trees? You mean our little tree will look like that someday?"*

*Michael looked at the big trees and tried to imagine the little tree growing up to be that big. They were so wide and tall. Michael suddenly realized that the big mulberry trees looked just like the shade trees in front of his house. He smiled.*

*That evening, Michael was waiting for his dad to come home from work. When he saw him coming, Michael ran to the door.*

*"Dad, the shade trees on our street are mulberry trees! Mrs. Yee showed me a big mulberry tree at school, and it is just like our trees."*

*Michael's dad looked at the trees. "You know what?" he asked. "I bet that's why they call this Mulberry Street!" Michael and his dad looked at each other and laughed.*

*After a minute Michael got an idea and said, "Hey, Dad. Let's go take a walk on Pine Street. I bet I know what kind of trees grow on that street!"*

Poster 4

**Materials for Step 2**
- *String*
- *Scissors*
- *Tree*

**TEACHING NOTE**

*Explain to students that* living *means alive and* nonliving *means not alive. People and trees are living, but rocks and soil are nonliving. People and trees are alive because they change and grow.*

2. **Introduce the class tree**

   At the end of the story, tell students that you have a surprise for them. Bring out the tree in its container. After students have had an opportunity to ask questions about the tree, allow time for them to observe the tree closely. Let them stand next to the tree to compare its height to theirs, use their fingers to see how big around the trunk is, measure the circumference with a string, and carefully feel the texture of the bark and leaves.

   Extend students' thinking by asking questions.

   ➤ *Do you think this tree is living or nonliving? Why do you think so?*

   ➤ *Do you think this tree is young or old? How do you know?*

   ➤ *How does the tree resemble its parent tree?*

   ➤ *What will our tree need while it stays in our classroom?*

3. **Discuss care and handling**

   After listening to students' ideas, confirm that the tree is living and that living things have basic needs. Discuss the need for soil for support, air, space, and nutrients. Discuss the need for water, but not too much water. Plan a watering schedule with students. Let them know that, while trees need sunlight, this tree will be all right in the classroom for a week or two without sunshine. Also tell students that it is OK to touch the tree but that they should do so gently, being careful not to pull any of the leaves or bark off the tree.

**Materials for Step 4**
- *Scrapbook*
- *Transparent tape*
- *White glue*
- *Camera*

4. **Add to class scrapbook**

   Tell students that they will be adding information about the tree to the class scrapbook. Take a picture of students with the class tree, date it, and paste it in the scrapbook. Also include the circumference string, any students' comments or observations, and one leaf from the tree.

5. **Identify a place for the tree in the classroom**

   Move the tree to a place in the room where it can stay for the next 2 weeks. If you are going to plant the tree, tell students that the tree will remain in the classroom until the tree-planting ceremony. They can be thinking about the things that can be included in the ceremony. Make plans to create invitations to send to families and school administrators, sing songs, and recite prose or poems.

**B R E A K P O I N T**

### 6. Plan tree-planting ceremony

Plan a ceremony to dedicate the tree to be planted. Turn the event into a special occasion. Invite the principal and students' families. As a class, practice songs and poetry recitations. Plan for each student to say a little piece about what kind of tree it is, how old it is, what the bark feels like, where it came from, how big it might get, what its leaves will look like, and so on.

### 7. Go outdoors

When all plans have been developed, take students outdoors to plant the tree. Be sure everyone gets to shovel some soil or take part in some other way. Discuss with students how the tree will have to be cared for, and remind them that they will be watching the tree grow over the years. Take pictures of the ceremony to add to the scrapbook.

**Materials for Step 7**
- *Planting tools*
- *Camera*
- *Water*

### 8. Return to class

Return to the classroom and have students wash their hands with warm, soapy water and clean up any tools that need washing.

### 9. Focus question: What do trees need to grow?

Write the focus question on the chart as you read it aloud.

➤ *What do trees need to grow?*

Distribute focus-question strips for students to glue into their notebooks.

Ask students to write or draw about what trees need to grow.

**EL NOTE**

*Provide a sentence frame for beginning writers, such as: A tree needs _____ to grow.*

## WRAP-UP

### 10. Share notebook entries

Conclude Part 6 by having students share notebook entries. Ask students to open their science notebooks to the most recent entry and pair up with a partner to

- share their answers to the focus question;
- explain their drawings.

**FOCUS CHART**

*What do trees need to grow?*

Trees need air, space, soil, water, and light.

# INTERDISCIPLINARY EXTENSIONS

## Language Extension

- **Make a tree-observation class book**
  Make a book that includes a drawing or painting of a tree from each student. If this is done soon after students complete their schoolyard walk to adopt a tree, use the students' observations that you recorded on the walk. Have students glue or paste their comments on a piece of paper and draw a picture to go with it. Designate a separate section for each adopted tree.

## Art Extension

- **Make more puzzles**
  Have students draw trees on card stock. Make puzzles by cutting each drawing into pieces. Have students share their puzzles with friends at school and with their families at home.

## Math Extensions

- **Hang up the trunk-circumference strings**
  Hang up the pictures of your schoolyard trees and the strings that represent their circumferences. Let students take the strings down and make them into circles to compare the tree sizes and remember how big the trunks are. Be sure to make a circumference string for your class tree.

- **Measure the circumference strings**
  Have students use centicubes, paper clips, or other arbitrary units of measure to determine the length of the circumference strings. Have them compare the strings from trees adopted at school and those at home.

## Science Extensions

- ### Plan a learning center

  Set up a table or special area in the classroom where you and your students will put things as you continue the study of trees. You can begin the center with the loupes/magnifying lenses and tree-trunk rounds. Leave plenty of space; this center will grow as the module continues.

  The book, *Trees*, provided in the kit, is a good reference for you and students to use at the center and throughout the module.

- ### Collect prunings

  If some trees are being trimmed or pruned in the vicinity of the school, go out with students to get some branches and bring them back to the classroom. Have students observe what happens to the leaves and bark. Talk about live and dead leaves.

## Home/School Connection

Students survey their family members to see how many trees each of them estimates to be in their yard, block, or local park. When all the predictions are in, students take a walk to make an actual count. Make copies of teacher master 13, *Home/School Connection* for Investigation 1, and send it home with students after Part 4.

### TEACHING NOTE

*Review the online activities for students on FOSSweb for module-specific science extensions.*

*No. 13—Teacher Master*

**Part 1**
Leaf Walk ............................ **100**

**Part 2**
Leaf Shapes ......................... **106**

**Part 3**
Comparing Leaves ............... **112**

**Part 4**
Matching Leaf Silhouettes .... **117**

**Part 5**
Leaf Books ......................... **122**

# PURPOSE

## Content

- Different kinds of trees have different leaves.

- Leaves have properties: size, shape, tip, edge, texture, and color.

- Leaf properties vary.

- Leaves can be described and compared by their properties.

## Scientific and Engineering Practices

- Observe the sizes, shapes, textures, and colors of tree leaves.

- Compare the shapes of leaves to common geometric shapes.

- Compare the sizes and edges of leaves.

- Record and communicate the similarities and differences among leaves.

# INVESTIGATION 2 – *Observing Leaves*

| Investigation Summary | Time | Focus Question |
|---|---|---|
| **Leaf Walk**<br>Students read about and discuss how we use our senses to learn. They take a schoolyard walk to observe leaves on trees, noting similarities and differences and gathering leaves to press and keep. | **Reading**<br>15 minutes<br>**Outdoors**<br>15 minutes<br>**Leaf Sorting**<br>20 minutes | What can we observe about leaves? |
| **Leaf Shapes**<br>Students look closely at the shapes of leaves and match leaves to geometric shapes. | **Center**<br>20 minutes<br>**Notebook**<br>15 minutes | What shapes are leaves? |
| **Comparing Leaves**<br>Students go outdoors for a leaf hunt. Using a paper reference leaf, they look for leaves that differ in size and shape. | **Outdoors**<br>15 minutes<br>**Notebook**<br>15 minutes | How are leaves different? |
| **Matching Leaf Silhouettes**<br>Students work in centers with representational materials to develop their skills of observation and comparison, matching leaf silhouettes and outlines. Leaf shape, size, and edges are the properties students use for comparisons. | **Center**<br>30 minutes | How are leaf edges different? |
| **Leaf Books**<br>Students make leaf books to add to their science notebooks. The teacher reads *Our Very Own Tree*, which summarizes many of the ways students have studied trees. | **Center**<br>30 minutes<br>**Reading**<br>15–20 minutes | What can we observe about leaves? |

PART 1 | PART 2 | PART 3 | PART 4 | PART 5

# At a Glance

| Content | Writing/Reading | Assessment |
|---|---|---|
| • Different kinds of trees have different leaves.<br>• Leaves have properties: size, shape, tip, edge, texture, and color. | **Science Notebook Entry**<br>Draw or write words to answer the focus question.<br>**Book**<br>*How Do We Learn?* | **Embedded Assessment**<br>Teacher observation |
| • Leaf properties vary.<br>• Leaves can be described and compared by their properties. | **Science Notebook Entry**<br>Draw or write words to answer the focus question. | **Embedded Assessment**<br>Teacher observation |
| • Leaves can be described and compared by their properties. | **Science Notebook Entry**<br>Draw or write words to answer the focus question. | **Embedded Assessment**<br>Teacher observation |
| • Leaves have properties: size, shape, tip, edge, texture, and color.<br>• Leaves can be described and compared by their properties. | Match the leaf silhouettes to answer the focus question verbally. | **Embedded Assessment**<br>Teacher observation |
| • Different kinds of trees have different leaves.<br>• Leaves have properties: size, shape, tip, edge, texture, and color.<br>• Leaves can be described and compared by their properties. | **Book**<br>*Our Very Own Tree* | **Embedded Assessment**<br>Teacher observation |

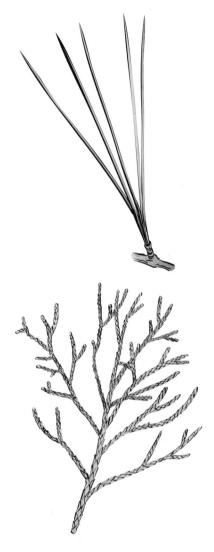

## BACKGROUND *for the Teacher*

When we try to identify a tree, we start with its leaves. Leaves have many **properties**. Leaves come in many shapes and **colors** and are important not only to the tree, but to countless other life-forms.

### What Can We Observe about Leaves?

Trees have a lot of leaves, and so trees provide shade. The shade is the result of the leaves performing their primary function—to intercept and absorb the Sun's rays. It is primarily the leaf that performs the wonder of transforming light energy into chemical energy, known generally as food.

When you wander under a tree and study the ground, in all likelihood you will see leaves that have fallen from the tree. Some fallen leaves are the result of unplanned events, such as violent wind, foraging animals, and falling objects. Other fallen leaves are part of the natural cycle of life events. Deciduous trees conduct a predictable end-of-summer ritual, starting with a change of leaf color and ending with complete defoliation of the tree. The entire burden of leaves drops, and the tree stands dormant until the next spring.

Evergreen trees shed leaves, too, but not all at once, so the continuous renewal process is less obvious. The careful observer will discover the discarded, worn leaves on the ground and notice the young, emergent leaves on the branches and twigs taking the place of the fallen leaves.

### What Shapes Are Leaves?

Leaves come in just about any shape you can imagine, including **spear**, **oval**, **heart**, **line**, **triangle**, and **paddle**. The main subdivision, which embraces most of the world's trees, is the clan called broadleaf (or hardwood) trees, and their very distant cousins the conifers. Broadleaf trees have broad, flat leaves in hundreds of wonderfully diverse patterns. Willow leaves are long, narrow and **pointed**, aspen leaves are **rounded**, linden leaves are heart-shaped, and the tulip-tree leaf is the shape of a T-shirt. Conifers, also called evergreens, have adapted to withstand the freezing winters of temperate and arctic climates. Their leaves are typically small and narrow, taking the form of the familiar needles of pine, spruce, and fir trees, and the less familiar overlapping scales of trees such as junipers and cedars. Conifer leaves may endure for 2 years or for as many as a dozen years before they are replaced.

## How Are Leaves Different?

The most obvious difference between leaves is their **size**. Early-childhood students will be able to find big leaves and little leaves to show you. With guidance, they will be able to compare specific dimensions of leaves to give "big" and "little" a bit more communication power. When given a reference leaf, students will be able to find leaves that are **longer** or **shorter**, **wider**, or **narrower**. More advanced observers will be able to compare the texture, thickness, and color of leaves as well.

To the experienced botanist, leaves differ in more complex ways. Leaf design is an adaptation that helps the bearer survive in its environment. The leaves of mesquite trees are tiny, tough, and covered with a waxy material. Small size and waterproofing help the tree resist intense heat and strong winds, which would desiccate and batter less well-fortified leaves. Contrast the mesquite with the dogwood tree that grows in the understory of temperate hardwood forests. The dogwood leaves are large, thin, intensely green (due to chlorophyll), and oriented parallel to Earth's surface. This leaf strategy allows the dogwood to capture the weak, filtered light that penetrates the forest canopy.

The leaves of redwood trees along the Northern California coast are adapted to intercept fog and concentrate the mist into substantial drops that fall to the ground near the trees' roots. A substantial portion of redwoods' required water is raked from the air by the featherlike leaves. The leaves of palm trees are long and flexible, designed to withstand persistent, often strong coastal winds. The compact leaves of alpine conifers are not only tough, but infused with resins that act as antifreeze to survive subzero temperatures.

**Smooth edge**

## How Are Leaf Edges Different?

The perimeter of a leaf (as observed in a leaf **outline** or **silhouette**) tells a lot about the leaf. It defines the shape and exhibits adaptations that communicate something about the life of the tree. The beginner sees that the **edge** is **smooth**, **toothed**, or **lobed**. A smooth margin is uninterrupted by protrusions or indentations. A toothed margin has a sawlike edge with regularly spaced zigzag points. A lobed margin is the most irregular edge, with extensions that radiate from a central location, like a classic maple leaf, and sinuses that isolate sections of the leaf, like a red oak leaf. Other leaves, such as holly and live oak leaves, have spines on the margins of the leaves. The shape of the edge lends interest to leaves and provides information for us to identify the tree. In addition, the shape of the edge may contribute to the tree's survival.

**Toothed edge**

**Lobed edge**

Different kinds of leaves are advantageous in different environments. Some leaves have asymmetrical blades, which tilt the leaves to one side and allow water to drain off quickly. Dumping water helps the plant dry after rain, reducing fungus growth and bacterial infection. A point on the outer edge also allows water to drain quickly. In warmer areas where water evaporates more quickly, sharp **tips** aren't necessary. Lobes increase the surface areas of leaves. This allows the leaves to transpire (give off) more water, escape damage from the wind, and take in carbon dioxide more efficiently.

**New Word**

Say it
See it
Hear it
Write it

*Color*
*Edge*
*Heart*
*Line*
*Lobed*
*Longer*
*Narrower*
*Outline*
*Oval*
*Paddle*
*Pointed*
*Property*
*Rough*
*Rounded*
*Shorter*
*Silhouette*
*Size*
*Smooth*
*Spear*
*Tip*
*Toothed*
*Triangle*
*Wider*

# TEACHING CHILDREN *about* *Observing Leaves*

Your role as science teacher in an early-childhood classroom is to guide observations, exposing students to as many real-world experiences as possible. The more experiences they have and the greater the variety, the more they are adding to their repertoire of knowledge, allowing them to see greater possibilities of how all things go together.

Students should be guided to look carefully at the structure of leaves, not only so they can tell that something is a leaf, but to appreciate the great diversity of nature. You'll be amazed at the tiny differences you and your students will become aware of. As students look more closely, they will have more questions, and their questions will guide them to find out more about the world around them.

The **conceptual flow** for this investigation starts with a leaf walk outdoors, during which students confirm that **trees have leaves**. They collect leaves from several trees and sort them to learn that **different kinds of trees have different kinds of leaves**.

In Part 2, students compare leaves to geometric shapes. They find that **leaves vary in shape** and can be sorted into groups based on shape: **line**, **spear**, **heart**, **oval**, **paddle**, and **triangle**.

In Part 3, students compare leaves to a reference leaf to find that **leaves vary in size**. Some leaves are **long**, others are **short**; some leaves are **wide**, and others are **narrow**.

In Part 4, students study leaf **edges** for similarities. They find that leaves might have **smooth**, **toothed**, or **lobed** edges.

In Part 5, students make leaf books to review the properties of leaves that can be used to identify the leaves. They read a book that reinforces leaf concepts.

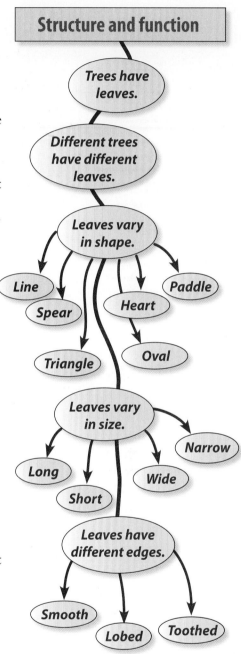

**FOCUS QUESTIONS A** ................................

| | |
|---|---|
| Inv. 1, Part 1: | What did we learn about our schoolyard trees? |
| Inv. 1, Part 2: | What are the parts of trees? |
| Inv. 1, Part 3: | What shapes are trees? |
| Inv. 1, Part 4: | Which trees have similar shapes? |
| Inv. 1, Part 5: | What can we find out about our adopted trees? |
| Inv. 1, Part 6: | What do trees need to grow? |
| Inv. 2, Part 1: | What can we observe about leaves? |
| Inv. 2, Part 2: | What shapes are leaves? |
| Inv. 2, Part 3: | How are leaves different? |
| Inv. 2, Part 4: | How are leaf edges different? |
| Inv. 2, Part 5: | What can we observe about leaves? |

FOSS Trees and Weather Module
© The Regents of the University of California
Can be duplicated for classroom or workshop use.

Investigations 1–2
No. 2—Teacher Master

*No. 2—Teacher Master*

## MATERIALS *for*

### Part 1: *Leaf Walk*

**For the class**

- 32   Zip bags, 1 L
- •   Self-stick notes
- •   Phone books or catalogs (from Investigation 1, Part 1) ★
- ❏   1   Teacher master 2, *Focus Questions A*
- 1   Book, *How Do We Learn?*

**For assessment**

- •   *Assessment Checklist*

★ Supplied by the teacher.      ❏ Use the duplication master to make copies.

# GETTING READY *for*
## Part 1: *Leaf Walk*

1. **Schedule the investigation**
   This is a whole-class activity. Plan 15 minutes to read *How Do We Learn?*, 15 minutes for a group walk outdoors, and 20 minutes for leaf sorting and notebook writing in the classroom.

2. **Preview Part 1**
   Students read about and discuss how we use our senses to learn. They take a schoolyard walk to observe leaves on trees, noting similarities and differences and gathering leaves to press and keep. The focus question is **What can we observe about leaves?**

3. **Select your outdoor site**
   Walk around your schoolyard and determine the route you will take with students when they look for leaves. Plan to visit several trees so students can find a variety of leaves. Try to include the following:

   - Trees with leaves that are starting to change color.
   - A broadleaf tree.
   - A tree with needles.
   - A tree with scales.
   - Trees with leaves arranged in different ways on the twigs.
   - Two trees of the same kind.

   Be sure to avoid areas where there are plants that students shouldn't touch (such as plants with thorns or poison oak or poison ivy).

4. **Check the site**
   It is always a good idea to check the outdoor site on the morning of an outdoor activity. Check for any distracting items or unsafe items where students will be working.

5. **Plan to read *How Do We Learn?***
   Plan to read the book *How Do We Learn?* to introduce this part (see Step 1 in Guiding the Investigation). This book introduces how we use our senses.

### 6. Plan assessment

There are six objectives that can be assessed during any part of this investigation.

**What to Look For**

- *Students ask questions.*

- *Students use their senses to observe living things.*

- *Students show respect for living things.*

- *Students record observations.*

- *Students communicate observations orally, in writing, and in drawings.*

- *Students use new vocabulary.*

Here are the content objectives for this part.

- *Leaves have properties that can be compared.*

- *Leaves can be described by their properties: size, shape, tip, edge, texture, and color.*

Focus on a few students each session. Record the date and a + or − on the *Assessment Checklist*.

# GUIDING *the Investigation*
## Part 1: *Leaf Walk*

1. **Read** *How Do We Learn?*
   Call students to the rug. Begin this investigation by reading *How Do We Learn?* This book introduces (or reinforces) how we use our five senses (see, hear, touch, smell, taste) to observe, and how we find out things by asking questions, reading, comparing, sorting, using tools, measuring, and making graphs.

2. **Prepare to go outdoors**
   Tell students that they will be going outdoors to visit their trees again. This time, they will look closely at leaves. They will look at, touch, and smell leaves. Remind students they should not taste anything they find outdoors.

3. **Go outdoors**
   Review the rules for venturing out. Travel to your first adopted tree in the usual orderly manner. Form a circle near the tree.

   Call for attention. Focus students' observations with questions.

   ➤ *Do you see any leaves?*

   ➤ *Where are the leaves? Point to them.*

   ➤ *What color are the leaves?* [In autumn and winter, leaves will show color variation.]

   ➤ *Are the leaves only on the branches and twigs?* [Some leaves will be on the ground.]

4. **Pick up leaves**
   Distribute a zip bag to each student. Ask each student to find one leaf on the ground that came from the tree. Students should put the leaves in their bags.

5. **Visit another tree**
   Move to a different kind of tree in the schoolyard. Ask each student to find a leaf from this second tree. Discuss the similarities and differences between the leaves from the two trees.

6. **Hunt for additional leaves**
   Have students go on a leaf scavenger hunt. Challenge them to find leaves from six different kinds of trees. Remind them to collect leaves from the ground rather than pulling them off a tree.

   Clearly describe the limits of the area in which students can roam to look for leaves. Commence the leaf hunt.

---

**FOCUS QUESTION**

*What can we observe about leaves?*

**Materials for Step 4**
• *Zip bags*

---

color
edge
lobed
pointed
rough
rounded
size
smooth
tip
toothed

7. **Return to class**

   When everyone has collected six different leaves, call students around, have them form a line, and return to the classroom in the usual orderly fashion.

8. **Focus question: What can we observe about leaves?**

   Ask the focus question.

   ➤ *What can we observe about leaves?*

   Write the focus question on the chart, and read it aloud. Distribute focus–question strips for students to glue into their notebooks.

9. **Find leaves that go together**

   Have students work in pairs to organize their leaves. Start by asking,

   ➤ *Are all the leaves the same **size** and shape?*

   Two students should put their leaves together and then sort them into groups of leaves that they think go together.

   For example, students might put leaves together that are the same **color**. Below are sorting properties you might suggest to students who are stalled.

   - Size (big or small).
   - Shape (round, oval).
   - **Edge** pattern (**smooth**, **toothed**, or **lobed**).
   - **Tip** shape (**rounded** or **pointed**).
   - Texture (**rough** or smooth).

10. **Review vocabulary**

    Visit the pairs as they match leaves and introduce or reinforce words that describe the properties of leaves. For example, leaf *edge* and leaf *tip* might be new terms for students. Add these words to the word wall.

### 11. Press the leaves

Bring out the phone books and catalogs you have collected for pressing leaves. Have each student lay their leaves flat on a page in one of the books. Label the pages with student names, using self-stick notes like index tabs.

### 12. Clean up

Set the books with leaves aside until they are needed later. Place something heavy on top of the books so the leaves will dry flat.

### 13. Answer the focus question

Have students use words and drawings to answer the focus question.

# WRAP-UP/WARM-UP

### 14. Share notebook entries

Conclude Part 1 or start Part 2 by having students share notebook entries. Ask students to open their science notebooks to the most recent entry. Read the focus question together.

➤ *What can we observe about leaves?*

Ask students to pair up with a partner to

*   share their answers to the focus question;

*   explain their drawings.

**Materials for Step 11**
*   *Phone books or catalogs*
*   *Self-stick notes*

---

**EL NOTE**

**Provide a sentence frame for beginning writers, such as: Some leaves are _____ .**

---

**FOCUS CHART**

What can we observe about leaves?

Some leaves are big. Some are small.

Some leaves are pointed.

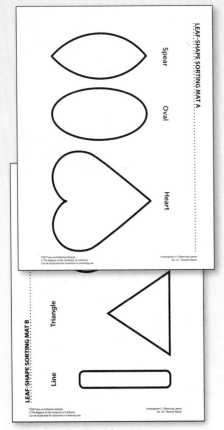

*No. 14—Teacher Master*

# MATERIALS *for*

## Part 2: *Leaf Shapes*

### For each student at the center

3–4 Leaves ★

### For each pair of students at the center

2 *Leaf-Shape Sorting Mats A* and *B* (See Step 7 of Getting Ready.)

### For the class

2 Sets of felt leaves, green, 9/set

1 Set of felt geometric shapes, yellow, 6/set

3 Zip bags

1 Small box or shopping bag ★

1 Felt board ★

5 Clear plastic sheet protectors (optional) ★

2–3 Basins or boxes to hold leaves ★

❑ 1 Teacher master 14, *Center Instructions—Leaf Shapes*

❑ 1 Teacher master 15, *Leaf-Shape Sorting Mat A*

❑ 1 Teacher master 16, *Leaf-Shape Sorting Mat B*

1 Teacher master 17, *Key to Leaf Names A*

### For assessment

• *Assessment Checklist*

★ Supplied by the teacher. ❑ Use the duplication master to make copies.

*Nos. 15–16—Teacher Masters*

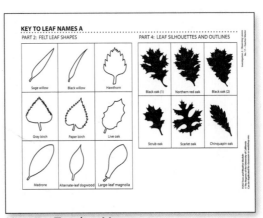

*No. 17—Teacher Master*

# GETTING READY *for*
## Part 2: *Leaf Shapes*

1. **Schedule the investigation**

   Work with groups of six to eight students at a center, or plan this part as a whole-class activity. Each group will need 15 minutes to work with the felt leaves and shapes and another 5–10 minutes to sort leaves. Plan 15 minutes for students to write or draw in their notebooks.

2. **Preview Part 2**

   Students look closely at the shapes of leaves and match leaves to geometric shapes. The focus question is **What shapes are leaves?**

3. **Inventory the felt leaves and shapes**

   There are two sets of felt leaves (9 leaves) and one set of felt geometric shapes (6 shapes). The six geometric shapes include spear, oval, heart, line, triangle, and paddle. Teacher master 17, *Key to Leaf Names A*, provides names for the nine felt leaf shapes.

   The set of geometric shapes and the two sets of leaf shapes are packaged in three separate zip bags. Inventory the felt leaves and geometric shapes to ensure that they are ready to use.

4. **Hide the felt shapes**

   Put the two sets of felt leaves and one set of felt geometric shapes in a box or shopping bag (still in their separate zip bags) to keep them out of view. The leaves and shapes will be brought out, one at a time.

5. **Set up the felt board**

   Use a felt board to introduce this part. If you do not have a felt board, you can use a bulletin board or cover a piece of cardboard with felt.

6. **Gather leaves**

   Each student will need three or four real leaves, each a different shape, to sort on copies of teacher masters 15 and 16, *Leaf-Shape Sorting Mats A* and *B*. Collect 50 or more leaves and needles of different kinds so that every student will have leaves representing several of the shapes on the mats. Put the leaves in two or three boxes or basins from which students will select them.

7. **Prepare the** *Leaf-Shape Sorting Mats*

Students work in pairs comparing real leaves to the outlines on the two *Leaf-Shape Sorting Mats*. Make a copy of both masters for each pair of students (e.g., if ten students will be working at a center, make five copies of both masters).

The *Leaf-Shape Sorting Mats* will be more durable if you insert them into clear plastic sheet protectors.

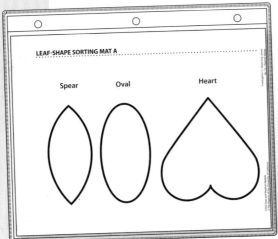
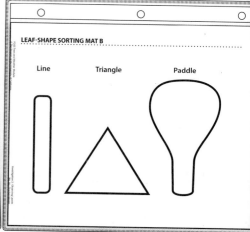

If the shapes of the leaves you collect for sorting are not well represented by the shapes provided, you can modify the sorting mats. For example, if you have maple leaves, you could include a star shape.

8. **Plan assessment**

Here are specific objectives for this part.

- *Leaves have properties that can be compared.*

- *Leaves can be described by their properties: size, shape, tip, edge, texture, and color.*

Focus on a few students each session. Record the date and a + or − on the *Assessment Checklist*.

# GUIDING *the Investigation*
## Part 2: *Leaf Shapes*

### 1. Introduce felt leaves

Bring your box of felt leaves and geometric shapes to the rug. Call students around. Place the nine leaf silhouettes (one set) on the felt board, one at a time. Compare similarities and differences as you put each new leaf silhouette on the board. Ask,

➤ *If these were real leaves, would they all come from the same tree?* [No.]

➤ *How can you tell?* [The leaves are all different shapes.]

### 2. Match the leaves

Place one felt leaf from the second set of felt leaves on the board. Ask students,

➤ *Is this leaf the same as any of the other leaves on the board?*

Move the leaf next to each of the felt leaves already on the board until students agree on a match.

Call on a student to come to the felt board, select a new felt leaf, find the leaf on the board that matches, and place the new leaf on top of the one it matches. Continue this process until all the leaves have been paired.

### 3. Review *shape*

Tell students,

*These leaves all have different shapes. Shape is a **property**. Shape is one way to tell if two leaves are the same or if they are different. We put these two leaves together because they have the same shape. Let's count together to see how many different shapes of leaves we have.*

Count together as you point to the nine different leaf shapes.

### 4. Introduce geometric shapes

Tell students that you have something else in the box. Take out the triangle and put it on the felt board. Ask students to name the shape. Ask,

➤ *Are there any felt leaves that are the same shape as the **triangle**?*

Have a student go to the felt board and choose a leaf that matches. Stick the felt leaf on top of the triangle shape. (There are two leaves that have triangle shapes, the gray birch and the hawthorn.)

*Trees and Weather Module*                                   **109**

---

**FOCUS QUESTION**

*What shapes are leaves?*

**Materials for Step 1**
- *Sets of green felt leaves*
- *Set of yellow felt shapes*
- *Box or bag*
- *Felt board*

Sage willow    Black willow    Hawthorn

Gray birch    Paper birch    Live oak

Madrone    Alternate-leaf dogwood    Large-leaf magnolia

**TEACHING NOTE**

*If the vocabulary for kinds of shapes is new, spend time introducing the words and writing them on the word wall.*

Say it · New Word · See it · Hear it · Write it

5.  **Observe other shapes**

    Tell students that you have other shapes in the box. One at a time, display the five additional shapes. Have a student choose a felt leaf to put on top of each shape. When finished, review the shapes of all the leaves.

    - *This leaf is **paddle**-shaped.*

    - *These two long leaves are **line**-shaped.*

    - *This pointed leaf is **spear**-shaped.*

    - *These rounded leaves are **oval**-shaped.*

    - *This leaf is **heart**-shaped.*

    - *These leaves are triangle-shaped.*

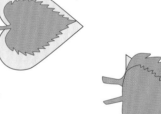

6.  **Introduce** *Leaf-Shape Sorting Mat*

    Tell students that now they are going to match real leaves to shapes. Show them the containers of leaves and the *Leaf-Shape Sorting Mats*. Assign partners to work together. Tell them that their job will be to look carefully at each shape, find a real leaf that goes with each shape, and place it on the matching shape on the mat.

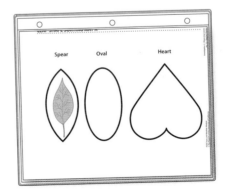

**7. Distribute materials**

Distribute copies of the *Leaf-Shape Sorting Mat*s and boxes or basins of leaves. Allow about 10 minutes for students to select three to six leaves and place them on the mats.

**8. Focus question: What shapes are leaves?**

After all students have had an opportunity to compare leaves and shapes, ask the focus question.

➤ *What shapes are leaves?*

Write the focus question on the chart as you read it aloud. Distribute focus-question strips for students to glue into their notebooks. Have students use words and drawings to answer the focus question.

**9. Answer the focus question**

Depending on students' experience with notebooks, you can let them work on their own, or you can model a notebook entry using the focus chart.

a. Ask students to choose one of the shapes (e.g., oval).

b. Draw an oval under the focus question.

c. Write "oval" under the drawing.

d. Draw a leaf in the drawing of an oval.

# WRAP-UP/WARM-UP

**10. Share notebook entries**

Conclude Part 2 or start Part 3 by having students share notebook entries. Ask students to open their science notebooks to the most recent entry. Read the focus question together.

➤ *What shapes are leaves?*

Ask students to pair up with a partner to

• share their answers to the focus question;

• explain their drawings.

**Materials for Step 7**

• *Leaf-Shape Sorting Mats*
• *Containers of leaves*

**EL NOTE**

**Provide a sentence frame for beginning writers, such as: This leaf is _____ .**

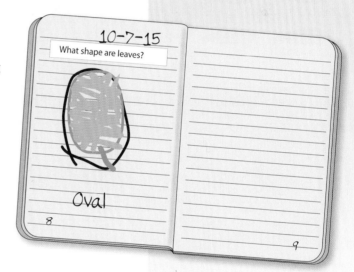

10-7-15

What shape are leaves?

Oval

8

9

**FOCUS CHART**

*What shapes are leaves?*

*Some leaves are oval.*

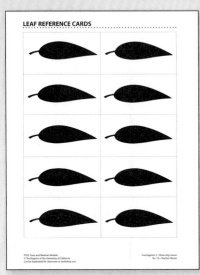

*No. 18—Teacher Master*

## MATERIALS *for*

### Part 3: *Comparing Leaves*

**For each student**

   1   Leaf reference card  (See Step 3 of Getting Ready.)

**For the class**

   1   Whistle or bell ★

   •   White glue ★

❏   1   Teacher master 18, *Leaf Reference Cards*

**For assessment**

   •   *Assessment Checklist*

★ Supplied by the teacher.      ❏ Use the duplication master to make copies.

# GETTING READY *for*

## Part 3: *Comparing Leaves*

1. **Schedule the investigation**

   This part is a whole-class activity. Plan 15 minutes for the outdoor activity and 15 minutes for notebook writing.

2. **Preview Part 3**

   Students go outdoors for a leaf hunt. Using a paper reference leaf, they look for leaves that differ in size and shape. The focus question is **How are leaves different?**

3. **Prepare the reference cards**

   Each student will need one image cut from teacher master 18, *Leaf Reference Cards*. Make copies of the duplication master accordingly.

4. **Select your outdoor site**

   Students will be hunting for leaves in the schoolyard. Determine the best area for students to find a variety of leaves. Limit the space you will allow students to explore. Establish a loud signal such as a whistle or bell to use when you want students to return to home base. Refer to the Taking FOSS Outdoors chapter (in *Teacher Resources*) for other strategies for working with students outdoors.

5. **Check the site**

   It is always a good idea to check the outdoor site on the morning of an outdoor activity. Check for any distracting items or unsafe items where the students will be collecting leaves.

6. **Plan assessment**

   Here are the objectives for this part.

   - *Leaves can be described and compared by their properties.*

   - *Compare properties of leaves.*

   Focus on a few students each session. Record the date and a + or − on the *Assessment Checklist*.

**Materials for Step 3**
- *Leaf reference cards*
- *Whistle or bell*

New Word
Say it
See it
Hear it
Write it

# **GUIDING** *the Investigation*
## Part 3: *Comparing Leaves*

1. **Review** *compare*

   Review the word *compare*. Students should know that it means to observe two things and find out how they are the same and how they are different.

2. **Explain the investigation**

   Tell students that today they are going on a leaf hunt outdoors. Review the rules and considerations for working outdoors.

   - Trees are living plants and need to be treated gently.
   - Look for leaves on the ground.
   - When the signal is sounded, come back to home base as quickly as possible.

3. **Go outdoors**

   Travel from the classroom to your home base in the usual orderly manner. Pick up a leaf on the way for a demonstration.

   Ask students to sit. Describe the boundaries within which students may hunt for leaves. Explain the signal you will use for attention.

4. **Demonstrate the first challenge**

   Show students the reference card from teacher master 18, *Leaf Reference Cards*. Tell them,

   *This card has a picture of a leaf. I'm going to see if I can find a leaf that is **longer** than the leaf picture on this card.*

   Hold up your demonstration leaf and the card. Ask,

   ➤ *Is the leaf I found longer than the picture of the leaf?*

   Students should agree that it is either longer or shorter than the image on the reference card.

5. **Distribute reference cards**

   Give each student a card for comparing leaves. Tell students,

   *Compare the size of this leaf to real leaves in the schoolyard. Find a leaf that is longer than the leaf on the card.*

   Send students on their way with a "Go." After 2 or 3 minutes, sound the signal for them to return. Have them share the leaves they found. Put them in one pile.

### 6. Provide more challenges

Repeat the procedure, sending students to find leaves that are

- **shorter** than the reference leaf;
- **wider** or **narrower** than the reference leaf.

If students get the size challenges quickly, ask them to find leaves that

- have edges that are different from the reference leaf. (The reference leaf has a smooth edge);
- are a different shape than the reference leaf. (The reference leaf is spear-shaped).

### 7. Return to class

Ask students to bring one leaf back to the classroom to put into their science notebooks. Collect the reference cards. Return to the classroom in the usual orderly manner.

### 8. Review vocabulary

Add to the word wall any words that came up in the outdoor activity. Be sure to include the words *longer*, *shorter*, *wider*, and *narrower*.

### 9. Focus question: How are leaves different?

Now that students have a number of experiences observing and comparing leaves, ask the focus question.

➤ *How are leaves different?*

Write the focus question on the chart and read it aloud. Distribute a focus-question strip for students to glue into their notebooks.

Visit each student and glue a reference card in each student's notebook under the focus question.

Have students glue their leaves into their notebook pages. Have each student write a word or a sentence that describes how his or her leaf is different from the reference leaf.

**New Word**
Say it
See it
Hear it
Write it

---

**EL NOTE**

*Tell students* **longer** *and* **shorter** *describe the distance from top to bottom of the leaf and* **wider** *and* **narrower** *describe the distance from one side of the leaf to the other (left to right).*

longer
narrower
property
shorter
wider

**Materials for Step 9**
- *White glue*

---

**EL NOTE**

*Provide a sentence frame for beginning writers, such as:* **This leaf is _____ .**

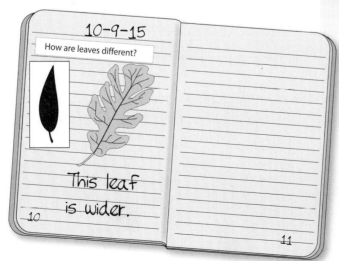

10-9-15

How are leaves different?

This leaf is wider.

10

11

## WRAP-UP/WARM-UP

### 10. Share notebook entries

Conclude Part 3 or start Part 4 by having students share notebook entries. Ask students to open their science notebooks to the most recent entry. Read the focus question together.

➤ *How are leaves different?*

Ask students to pair up with a partner to

* share their answers to the focus question;
* explain their drawings.

---

**FOCUS CHART**

*How are leaves different?*

*Leaves are different sizes.*

*Leaves have different edges.*

---

# MATERIALS *for*
## Part 4: *Matching Leaf Silhouettes*

### For each pair of students at the center

1   Set of leaf silhouettes and outlines (12/set)

1   Set of same-size leaf silhouettes (6 cards plus a strip of 6 leaves)

1   Set of big and little leaf silhouettes (6 cards plus a strip of 6 leaves)

### For the class

1   Teacher master 17, *Key to Leaf Names A*

❏   1   Teacher master 19, *Center Instructions—Matching Leaf Silhouettes A*

❏   1   Teacher master 20, *Center Instructions—Matching Leaf Silhouettes B*

1   Teacher master 21, *Key to Leaf Names B*

❏   1   Teacher masters 33–37, replacement masters for leaf silhouettes and outlines (optional)

### For assessment

• *Assessment Checklist*

❏ Use the duplication master to make copies.

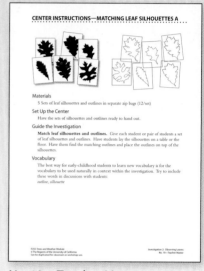

*No. 19—Teacher Master*

*No. 20—Teacher Master*

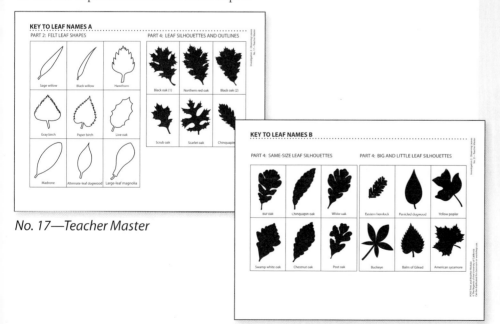

*No. 17—Teacher Master*

*No. 21—Teacher Master*

## GETTING READY *for*
### Part 4: *Matching Leaf Silhouettes*

1. **Schedule the investigation**
   This part involves pairs of students working with three different sets of leaf–silhouette matching materials at a center (allow 10 minutes per set). There are five copies of each set of matching materials— enough sets for 15 pairs of students working at the same time.

2. **Preview Part 4**
   Students work in centers with representational materials to develop their skills of observation and comparison, matching leaf silhouettes and outlines. Leaf shape, size, and edges are the properties students use for comparisons. The focus question is **How are leaf edges different?**

3. **Review the three sets of matching materials**
   Inventory the three kinds of leaf-silhouette materials. There are five sets of each kind. Refer to teacher masters 17 and 21, *Key to Leaf Names A* and *B*, for the names of each leaf shown as a silhouette or outline.

   - Leaf silhouettes and outlines consists of six silhouette cards (different kinds of oak leaves) and six transparent outlines of the same leaves. Students lay the silhouettes on a table and lay the matching outlines on top of the silhouettes.

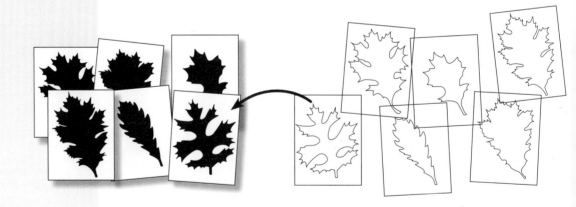

- Big and little leaf silhouettes consists of six different kinds of leaves on a strip of laminated blue paper and six matching cards that are slightly smaller. Students play two games using these leaf silhouettes—What Matches? and What's Missing?

- Same-size leaf silhouettes consists of six oak-leaf silhouettes on a strip of laminated green paper and six matching cards. Students play What Matches? and What's Missing? with these oak leaves.

**TEACHING NOTE**

*The same-size leaf silhouette cards have leaves with fine differences. When one card is missing, students must look closely at the fine detail to find out which one is missing.*

4. **Plan assessment**

   Here are specific objectives for this part.

   - *Leaves have properties: size, shape, tip, edge, texture, and color.*

   - *Leaves can be described and compared by their properties.*

   Focus on a few students each session. Record the date and a + or − on the *Assessment Checklist*.

**FOCUS QUESTION**
*How are leaf edges different?*

**Materials for Steps 1–3**
* *Leaf silhouettes and outlines*

**EL NOTE**
*Add the bold words to the word wall with a simple illustration.*

## GUIDING *the Investigation*
### Part 4: *Matching Leaf Silhouettes*

1. **Focus question: How are leaf edges different?**
   Call students to the rug. Review the different kinds of leaves students found in Part 3. Ask the focus question.

   ➤ *How are leaf edges different?*

   Write the focus question on the chart as you read it aloud.

2. **Review** *edge*
   Hold up two or three leaf-silhouette cards that have lobed edges. Use them to review leaf edge.

   **Examples of lobed leaves**

   Say,

   *Look at the edges of these leaves. When the edge of a leaf has parts sticking out like fingers, we say that the leaf has lobes. The lobes can be small or large. They can be pointed or rounded. The edges on these leaves are all lobed.*

   Ask the class if they see lobes on the leaf.

   ➤ *Are these leaves with lobes all the same?* [No.]

   *Today we will look at a lot of leaf* **silhouettes***. A silhouette shows the shape of the edge of a leaf.*

3. **Demonstrate silhouettes and outlines**
   Hold up one of the silhouettes and its matching outline. Ask,

   ➤ *How are silhouettes different from* **outlines***?* [Silhouettes are filled in with black, and outlines aren't.]

   Let students know that they will be challenged to find the matching outlines for each of the silhouettes when they are at the center. Demonstrate how to place an outline on top of the silhouette that it matches.

## 4. Demonstrate What Matches?

Show students the two different leaf strips and their corresponding cards (big and little leaf silhouettes and same-size leaf silhouettes). Demonstrate how to play What Matches? by setting out a leaf strip and the matching cards in random order. While students watch, find the card that matches one of the leaf silhouettes on the strip and place the card on top of the strip.

## 5. Demonstrate What's Missing?

Demonstrate how to play What's Missing? Set out a leaf strip and the matching cards in random order. Close your eyes while a student turns one of the cards face down. When you figure out which card is face down, point to the image of that card on the strip.

## 6. Send students to centers

The three different sets of matching materials can accommodate up to 15 pairs of students working simultaneously at three centers. Send students to the centers.

Visit students while they work. Ask questions to focus attention on specific properties of the leaf silhouettes.

➤ *Some leaves have smooth edges. Do any of these leaf silhouettes have smooth edges?*

➤ *Some leaves have toothed edges. Toothed edges look like a saw. Do you see any toothed leaves?*

➤ *Some leaves have lobed edges. Do any of the leaves have lobes on their edges?*

➤ *Are the lobes smooth or pointed?*

➤ *Where is the tip of the leaf?* [On the end opposite the stem.]

➤ *Is the tip the same on all the leaves?* [No.]

## 7. Review vocabulary

When students have worked with all three sets of matching materials, call them to the rug. Review the words that describe different edges of leaves along with the new words.

**Materials for Step 4**
- *Big and little leaf silhouettes*
- *Same-size leaf silhouettes*

Smooth edge

Lobed edge

Toothed edge

**TEACHING NOTE**

See the **Home/School Connection** for Investigation 2 at the end of the Interdisciplinary Extensions section. This is a good time to send it home with students.

edge
lobed
outline
pointed
silhouette
smooth
toothed

## MATERIALS *for*

### Part 5: *Leaf Books*

For each student at the center

1   Half sheet of white paper (8.5" × 11" cut in half) ★

For the class

* White glue ★
* Pressed leaves (from Part 1)

1   Book, *Our Very Own Tree*

For assessment

* *Assessment Checklist*

★ Supplied by the teacher.

# GETTING READY *for*

## Part 5: *Leaf Books*

1. **Schedule the investigation**
   Groups of six to ten students will need 20–30 minutes at a center. Plan 15–20 minutes to read *Our Very Own Tree*.

2. **Preview Part 5**
   Students make leaf books to add to their science notebooks. The teacher reads *Our Very Own Tree*, which summarizes many of the ways students have studied trees. Students revisit the focus question from Part 1, **What can we observe about leaves?**

3. **Plan leaf books**
   Cut standard 8.5" × 11" paper in half. Each student will use a folded half sheet. Students will glue one of their pressed leaves onto each of the four faces of the folded sheet.

4. **Plan to read *Our Very Own Tree***
   Read through the big book *Our Very Own Tree* with an eye for those experiences in the book that your class has shared.

5. **Plan assessment**
   Here are specific objectives for this part.

   - *Different kinds of trees have different leaves.*

   - *Leaves have properties: size, shape, tip, edge, texture, and color.*

   - *Leaves can be described and compared by their properties.*

   - *Students communicate orally, in writing, and in drawings.*

   Focus on a few students each session. Record the date and a + or − on the *Assessment Checklist*.

*What can we observe about leaves?*

**Materials for Steps 1–2**
- *Pressed leaves in books*
- *Half sheets of paper*
- *White glue*

# GUIDING *the Investigation*
## Part 5: *Leaf Books*

1. **Compare leaves**

   At the center, bring out the phone books and catalogs with students' pressed leaves. Help each student choose two leaves that look like they are from different trees. Have students take turns telling the group one or two ways that the leaves are different and one way that they are the same.

2. **Begin making books**

   Use the pressed leaves to make leaf books. Give each student a half sheet of paper. Show them how to fold it in half. Have them choose two more leaves and glue each leaf on a different face of the folded paper.

3. **Focus question: What can we observe about leaves?**

   Revisit the focus question.

   ➤ *What can we observe about leaves?*

   Write the focus question on the chart, and read it aloud. Distribute focus-question strips for students to glue into their notebooks.

4. **Add descriptions to the pages**

   Have students use the word wall and focus chart to write or dictate a word or phrase that describes each of their leaves and answers the focus question.

5. **Insert the leaf books**

   When the glue has dried, plan to insert the four-page leaf books into students' notebooks.

   a. *Open the notebook to the insertion location.*

   b. *Run a thin line or several dots of white glue in the gutter of the notebook.*

   c. *Insert the leaf book and close the notebook.*

6. **Read** *Our Very Own Tree*

Introduce the story *Our Very Own Tree*. Tell students that two girls have adopted a tree and that they will talk about what makes their tree special to them.

Read the story aloud. Pause and discuss the girls' experiences that were similar to the experiences students had with their adopted trees: feeling the trunk, looking at branches and twigs, and drawing pictures of leaves.

*Page 4*      *Does the bark on your tree feel rough?*

*Page 5*      *Do you remember looking at the roots and leaves of your tree?*

*Page 6*      *If the tree has acorns, what kind of tree is it?*

*Page 8*      *What else are acorns used for?*

*Page 11*     *Is your tree a house for animals?*

*Page 12*     *Does your tree have scars?*

7. **Discuss the story**

Discuss the story, using these questions as a guide.

➤ *What made it the girls' tree?*

➤ *Have you enjoyed spending time with your tree? How?*

➤ *Finish this sentence: Trees are special because _____ .*

➤ *Some worms are harmful. How are they harmful?*

# INTERDISCIPLINARY EXTENSIONS

## Language Extension

- **Make invitations**

    Have students use some of their pressed leaves to make invitations for back-to-school night or open house. Have students fold a piece of 14 × 21.5 cm (5.5" × 8.5") card stock in half, glue leaves on the front, and glue a written invitation on the inside.

## Math Extension

- **Make a leaf-shape bar graph**

    Draw a large grid on a piece of chart paper. Label each column with a geometric shape. Include an "Other" column for shapes that don't fit the ones studied in this investigation. As students bring in leaves, glue them to the graph to see which shape is the most common.

### TEACHING NOTE

*Refer to the teacher resources on FOSSweb for a list of appropriate trade books that relate to this module.*

## Art Extensions

- **Make leaf rubbings**

  Have students make rubbings of pressed leaves they have collected (in the same way that they made rubbings of bark in Investigation 1.)

- **Try spatter painting**

  Have students place a leaf on a piece of construction paper and use a toothbrush to spatter paint over a screen held above the construction paper and leaf. Remove the leaf, and a silhouette of the leaf remains.

- **Make photo frames**

  Use the spatter–painting technique to make photo frames. Have students cover the area where the photo will go with a piece of paper. Then have students put leaves around the edges and spatter paint. Paste the photo in the middle.

- **Make sunprints**

  Have students make sunprints.

  a. *Working in the shade, use a couple of small pieces of tape to secure a sheet of commercially available sunprint paper to a piece of cardboard.*

  b. *Arrange one or more leaves on the sunprint paper.*

  c. *Put a sheet of clear plastic on top of the arrangement to hold it in place.*

  d. *Expose the paper to direct sunlight for 5 minutes or so.*

  e. *Plunge the exposed sunprint paper in a basin of water. Dry the paper. You will have permanent silhouettes of the leaves.*

## Science Extensions

- **Add to the learning center**
  As the module continues, bring in leaves, branches, and other parts of trees for students to observe closely in the classroom. Have students add things they bring from home to share with their classmates.

- **Continue matching challenges**
  Photocopy teacher masters 15 and 16, *Leaf-Shapes Sorting Mat A* and *B*, and cut out the shapes. Have students glue leftover leaves from other projects to the corresponding shapes. Use as many of the geometric shapes as you can find leaves to match. Glue the shapes and leaves on a page of the scrapbook.

- **Use a two-handed feely box**
  Cut holes on two panels of a box at least 30 cm (1') on a side. Put a set of leaves (with gross differences) into the box. Give students a leaf that is shaped like one of the leaves in the box. Students try to find the leaf that matches.

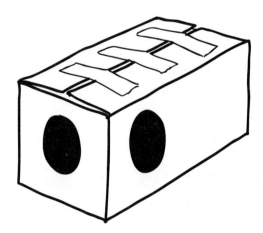

- **Extend What Matches? activity**
  Use the leaf-silhouette cards. Have partners begin together, standing on one side of the room with the reference strip in hand and with the cards laid out on a table on the other side of the room. Have partner A point to a leaf. Partner B's challenge is to walk across the room and bring back the card that matches that leaf.

## Home/School Connection

Students play Is This My Leaf? with their families to practice leaf identification.

Make copies of teacher master 22, *Home/School Connection* for Investigation 2, and send it home with students after Part 4.

*No. 22—Teacher Master*

**Part 1**
Weather Calendar ................ **138**

**Part 2**
Recording Temperature ........ **145**

**Part 3**
Wind Direction ................... **151**

# PURPOSE

## Content

- Weather is the condition of the air outdoors.
- Weather can be described as sunny, partly cloudy, overcast, rainy, or snowy.
- Temperature is how hot or cold it is.
- Thermometers measure temperature.
- Air temperature tells something about the weather.
- Wind is moving air.
- A wind sock indicates wind direction and speed.
- Weather changes.
- The Sun, Moon, and clouds are objects we see in the sky.

## Scientific and Engineering Practices

- Observe weather by using senses and simple tools.
- Compare weather from day to day.
- Record weather observations using pictures and words.

| Investigation Summary | Time | Focus Question |
|---|---|---|
| **Weather Calendar**<br>Students share what they know about weather and how it relates to air. A class weather monitor begins recording daily weather observations on a class calendar. Weather pictures (symbols) are used to indicate five basic types of weather. | **Outdoors**<br>15 minutes<br><br>**Calendar and Notebook**<br>20 minutes<br><br>**Reading**<br>15 minutes | **What is the weather today?** |
| **Recording Temperature**<br>Students use a thermometer and take turns measuring and recording the relative temperature (freezing, cold, cool, warm, hot). | **Introduction**<br>15 minutes<br><br>**Outdoors**<br>15 minutes<br><br>**Calendar and Notebook**<br>15 minutes | **How can we measure the air temperature?** |
| **Wind Direction**<br>Students construct a wind sock and observe how it responds when air moves through it. They find out that they can determine wind direction by using a wind sock. | **Introduction**<br>15 minutes<br><br>**Making socks**<br>15–20 minutes<br><br>**Outdoors**<br>10 minutes<br><br>**Notebook**<br>10 minutes<br><br>**Reading**<br>15 minutes | **What does a wind sock tell us about the wind?** |

PART 1

PART 2

PART 3

# At a Glance

| Content | Writing/Reading | Assessment |
|---|---|---|
| • Weather is the condition of the air outdoors.<br>• Weather can be described as sunny, partly cloudy, overcast, rainy, or snowy.<br>• Weather changes.<br>• The Sun, Moon, and clouds are objects we see in the sky. | **Science Notebook Entry**<br>Draw or write words to answer the focus question.<br>**Science Resources Book**<br>"Up in the Sky" | **Embedded Assessment**<br>Teacher observation |
| • Temperature is how hot or cold it is.<br>• Thermometers measure temperature.<br>• Air temperature tells something about the weather. | **Science Notebook Entry**<br>Draw or write words to answer the focus question. | **Embedded Assessment**<br>Teacher observation |
| • Wind is moving air.<br>• A wind sock indicates wind direction and speed. | **Science Notebook Entry**<br>Draw or write words to answer the focus question.<br>**Science Resources Book**<br>"Weather" | **Embedded Assessment**<br>Teacher observation |

# BACKGROUND *for the Teacher*

You've heard it before—everyone talks about the **weather**, but no one does anything about it. It's very difficult to do anything about something that few people really understand.

Weather is very complex. The word *weather* comes from an ancient language spoken over 5000 years ago; it is derived from *we,* meaning wind, and *vydra,* meaning storm. Weather is more than windstorms, but weather is always directly or indirectly something going on with or in the **air**. Weather is defined as the condition of the atmosphere at a point in time. Weather conditions include **temperature** (how **hot** or **cold** the air is), humidity and **clouds** (how much water is in or falling out of the air), **wind** speed and **direction** (speed and direction of **moving air**), and air pressure (what the density of the air is).

Weather is a product of complicated interactions among the gases that constitute the air; heat from the Sun; and water, both in the ocean and in the atmosphere. The Sun is primarily responsible for making weather happen on our planet. The energy from the Sun, influenced by Earth's rotation, the tilt of Earth's axis, and the arrangement of land and water, produces the uneven heating that drives Earth's weather. This is heady stuff—mostly inaccessible to early-childhood students. However, our budding meteorologists can make some basic weather observations and report their experiences in a scientific manner.

## What Kind of Weather Are We Having Today?

Our sensory systems provide empirical evidence about the weather. We feel hot and cold through the temperature-sensing nerve endings in our skin, we feel the wind buffeting the pressure-sensing nerve endings in our skin, and some folks claim they can smell rain approaching. Most of us simply use our visual sensory system to take a look out the window or up into the sky to make a judgment about the weather. A quick look informs us that it is clear and **sunny, overcast**, foggy, **rainy, snowy**, windy, calm, or some combination of these conditions.

The first thoughtful engagement with weather is recognizing its key indicators. Clouds, wind, fog, rain, snow, sunshine, heat, and cold all carry information. The second engagement with weather is keeping track of the key indicators of weather, day by day. Is it **partly cloudy** again today? Yes, so let's draw some clouds on the **calendar**. A record allows young students to look for patterns. A pattern of weather starts students on the path of understanding one more element of the way their world works.

A subtext in this investigation is awareness of objects and activities in the sky. We turn our gaze skyward to check the weather and not infrequently spy something else of interest. The predictable, recurring sightings include the Sun, Moon, and stars (including planets). These luminous or reflective objects appear in the sky with mathematical regularity. Less predictable, but equally familiar, is the endless variety of clouds. Add in the occasional bolt of lightning, and that about sums up the naturally occurring things students are likely to see in the sky.

Another set of stuff seen in the sky comprises earthbound objects that are swept aloft by wind. The two classes of airborne detritus include organic bits and pieces, such as leaves and feathers, and inorganic materials in the form of dust (clay, silt, and sand particles).

Students will be familiar with a third class of objects observed in the sky—living organisms. Birds and insects have mastered flight, giving them access to the skies. Students may tell you that they see trees in the sky.

Finally, students may sight any number of ingenious human artifacts in the sky. Powered contraptions such as airplanes and blimps rumble and thunder across the sky. Humans also fill the air with passive contrivances such as parachutes, kites, and balloons. The careful observer may spot one or more of the thousands of artificial satellites circling Earth, which can be seen on a clear night just after sunset or just before sunrise.

## How Can We Measure the Air Temperature?

Weather has four major components that can be **monitored** with **weather instruments** and continually concern people: the amount of motion of the air, the atmospheric pressure, the amount of water in or precipitation from the air, and the amount of heat energy in the air. This fourth component, heat, is monitored as temperature—the average kinetic energy of the gas particles in the atmosphere. To monitor temperature, we reach for a **thermometer**. A thermometer is a device that responds to the kinetic energy of the air and provides a reading on a scale.

Early-childhood students will not understand how a thermometer works, but they can see it change and compare the change to the observed environmental conditions. The column of liquid will go down when the thermometer is placed in cold water; the column of liquid will go up when the thermometer is placed in warm water. Various levels of the liquid column can be equated with degrees of warmth. If the column is down there, the weather is cold. If the column is way up there, it is really hot. A simulated thermometer is useful for orienting young temperature monitors to the appearance of the thermometer column and the temperature that the column communicates.

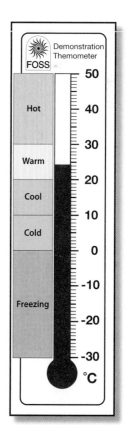

**Demonstration thermometer**

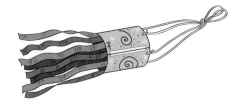

## What Does a Wind Sock Tell Us about the Wind?

Wind is matter (air) in motion. Matter in motion has momentum. When matter in motion encounters another chunk of matter, a force is applied. Force applied to matter changes motion. Things change speed and/or change direction. When wind hits a wind sock, it moves. The way the wind sock moves conveys information to the observer.

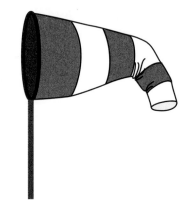

A classic wind sock is a truncated fabric cone, open at both ends. The larger end is held open by a hoop that is mounted on a shaft that is free to rotate. When the force of the wind impacts the wind sock, it first acts like a rudder, rotating the wind sock so that the larger opening faces into the wind. The wind entering the sock then impacts the inside walls, pushing them out of the way. If the breeze is stiff, the entire sock will be inflated, and it will be fully extended. If the breeze is modest, the weight of the fabric will cause some portion of the cone to flop down. Thus, the direction the wind sock faces indicates the wind direction, and the degree of inflation indicates the force (speed) of the wind. Simple and ingenious.

Early-childhood students can make a simple **wind sock** with a paper cylinder with **streamers** trailing out from the back. When students troop out to the schoolyard proudly holding their freshly constructed wind socks, they will feel the wind on their cheeks and on their simple meteorological instruments. There will be a connection. They might think that the wind that blows dust in their faces and pulls at their scarves is also pushing on their wind socks and everything else in its path. Trees, flags, loose papers, and dead leaves are all animated by the wind. Students can read it all around them. The wind sock becomes a tool for acquiring information about wind without students actually feeling the wind in their faces.

Say it
See it
Hear it
Write it
**New Word**

Air
Blowing
Calendar
Cloud
Cold
Cool
Direction
Freezing
Hot
Monitor
Moving air
Overcast
Partly cloudy
Rainy
Snowy
Streamer
Sunny
Temperature
Thermometer
Warm
Weather
Weather instrument
Wind
Wind sock

# TEACHING CHILDREN *about* *Observing Weather*

Early-childhood students come to the **Trees and Weather Module** with a substantial quantity of information about weather. They know when it is hot and cold, when it is windy, rainy, and snowy. They may even have had some exposure to violent and destructive weather (thunderstorms, tornadoes, hurricanes, etc.), either through personal experiences or via the media. You may find that students already have a rich and colorful vocabulary to describe both mundane and extreme weather conditions.

Even though students will be observing different kinds of weather and measuring the tangible components of weather, such as temperature and wind direction, there is no effort to explain the origin of these phenomena in terms of atmospheric physics, the water cycle, and climatology. That can come later. For now, it is sufficient for students to become familiar with the components of weather and the changes in weather in their area and to use some of the tools that help monitor and measure weather phenomena.

The **conceptual flow** starts with a discussion about **weather, the condition of the air**. Students trek outdoors to observe the weather, using their senses. After a few minutes, students **describe** the weather as **sunny**, **partly cloudy**, **overcast**, **rainy**, or **snowy**. Students learn how to record the weather, using **pictures**, **words**, and a **calendar**.

In Part 2, students learn that a **thermometer is used to measure temperature**. After observing that warm water makes the red line go up, and cold water make the red line go down, students measure the temperature outdoors. They report the temperature as being **freezing**, **cold**, **cool**, **warm**, or **hot**.

In Part 3, students make a **wind sock** and use it to **measure wind**. They learn that the wind sock can indicate wind **direction** and wind **speed**.

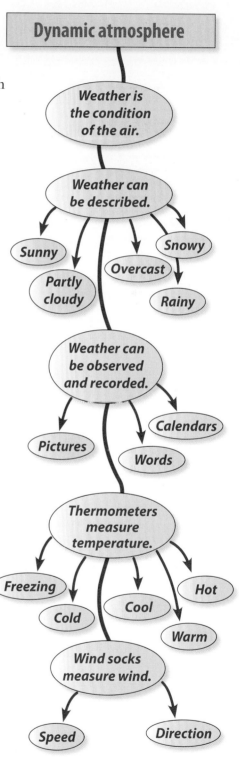

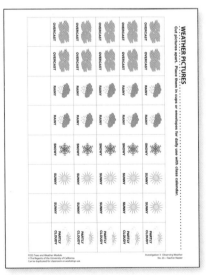

*No. 3—Teacher Master*

## MATERIALS *for*
### Part 1: *Weather Calendar*

### For each student

1 *FOSS Science Resources: Trees and Weather*
- "Up in the Sky"

### For the class

1 Class calendar

1 Watercolor pen or overhead-transparency pen ★

5–6 Cups or envelopes ★

1 Scissors ★

• Transparent tape ★

❏ 1 Teacher master 3, *Focus Questions B*

❏ 1 Teacher master 23, *Weather Pictures*

1 Big Book, *FOSS Science Resources: Trees and Weather*

### For assessment

• *Assessment Checklist*

★ Supplied by the teacher.     ❏ Use the duplication master to make copies.

*No. 23—Teacher Master*

# GETTING READY *for*
## Part 1: *Weather Calendar*

1. **Schedule the investigation**

   This is a whole-class activity. Plan on 15 minutes outdoors to observe weather, 20 minutes indoors to introduce the class weather calendar and for notebook writing, and 15 minutes for reading.

2. **Preview Part 1**

   Students share what they know about weather and how it relates to air. A class weather monitor begins recording daily weather observations on a class calendar. Weather pictures (symbols) are used to indicate five basic types of weather. The focus question is **What is the weather today?**

3. **Prepare the weather calendar**

   Your class calendar has no numbers. Decide if you want to write the dates on the calendar ahead of time or with students as part of the learning experience. You will need a watercolor pen or overhead-transparency pen to write on the laminated calendar.

4. **Prepare weather pictures for the calendar**

   Students will record the weather by sticking pictures for rain, snow, sun, overcast, and partly cloudy on the weather calendar. Make copies of teacher master 23, *Weather Pictures,* as needed.

Students may want to create other weather pictures, such as fog, thunderstorm, hail, and so on. Decide on a standard picture that will be used and include it on the bottom row of the *Weather Pictures* master sheet before making copies.

Cut the pictures apart and sort them. Place each type of weather picture in a cup or envelope. Keep these near the calendar for the class weather monitor to choose from.

**TEACHING NOTE**

*If possible, start this investigation on a day when there is a breeze or wind blowing and clouds in the sky. Plan a subsequent session for outdoor observations on any day when the weather changes and students can observe different weather conditions.*

| Sunday | Monday | Tuesday | Wednesday | Thursday | Friday | Saturday |
|--------|--------|---------|-----------|----------|--------|----------|
|        |        |         |           |          |        |          |
|        |        |         |           |          |        |          |
|        |        |         |           |          |        |          |
|        |        |         |           |          |        |          |
|        |        |         |           |          |        |          |
|        |        |         |           |          |        |          |

OVERCAST    RAINY    SUNNY    SNOWY    PARTLY CLOUDY

5. **Plan for weather monitors**

   Assign one student to be weather monitor each day. His or her task will be to record weather observations on the class calendar for that day. The student should choose the weather picture that most closely matches the weather and tape it to the class calendar.

6. **Select your outdoor site**

   In this part, students go outdoors briefly to check the day's weather conditions. Plan to take the class where they can line up, such as along a building wall or a line painted on the playground. If the weather is too wet or cold to go out, plan to gather the class at a window to discuss the weather conditions that they are able to see.

7. **Check the site**

   It is always a good idea to check the outdoor site on the morning of an outdoor activity. Check for any distracting items or unsafe items where students will be working.

8. **Plan to read *Science Resources*: "Up in the Sky"**

   Plan to read "Up in the Sky" during a reading period near the end of this part.

9. **Plan assessment**

   There are several objectives that can be assessed at any time during any part of this investigation.

   **What to Look For**

   - *Students ask questions.*

   - *Students record and organize observations.*

   - *Students communicate observations orally, in writing, and in drawings.*

   - *Students use new vocabulary.*

   Here are specific objectives to observe in this part.

   - *Observe and describe objects in the sky.*

   - *Record weather observations, using pictures and words.*

   Focus on a few students each session. Record the date and a + or − on the *Assessment Checklist*.

# GUIDING *the Investigation*
## Part 1: *Weather Calendar*

1. **Connect air and weather**

   Call students to the rug. Tell them they are going to study weather along with studying trees. Ask,

   ➤ *Where do you find* **weather***?*

   ➤ *What does* **air** *have to do with weather?*

   Listen to students' answers and encourage responses that relate air and weather. Tell students that they will be going outdoors to feel the air and to check the weather.

2. **Go outdoors**

   Follow your procedures for leaving the classroom, and conduct an orderly transition to your outdoor home base. Form a sharing circle. Have students close their eyes and try to feel the air.

   After a minute, have students open their eyes to look at the sky. If the sun is out, caution students not to look directly at it. Ask,

   ➤ *How does the air feel?*

   ➤ *Can you feel the air moving?*

   ➤ *Do you see anything that tells you that the air is moving?*

   ➤ *Do you see* **clouds** *in the sky? Are the clouds moving?*

   ➤ *What kind of weather do you see and feel?*

   Tell students,

   *When people talk about the conditions of the air outdoors, they are usually talking about weather.*

3. **Return to class**

   Return to the classroom. Gather students at the rug. Ask,

   ➤ *What are some words we use to describe the weather, not just today but on any day?*

   As students offer words or phrases, record them on the word wall.

4. **Introduce weather pictures**

   Explain that one way to record information about the weather is with pictures. Show students the pictures that you cut apart from teacher master 23, *Weather Pictures*. Have students help you read the word on each picture. Tell them,

**FOCUS QUESTION**

*What is the weather today?*

**New Word**
Say it → See it → Hear it → Write it

**Materials for Step 5**
- *Class calendar*
- *Watercolor pen*
- *Weather pictures*
- *Transparent tape*

**October**

| Sunday | Monday | Tuesday | Wednesday | Thursday | Friday | Saturday |
|--------|--------|---------|-----------|----------|--------|----------|
| 1 ☀ | 2 ☀ | 3 ☀ | 4 ☁ | 5 ☁ | 6 ☁ | 7 ☀ |
| 8 ☀ | 9 ☁ | 10 | 11 | 12 | 13 | 14 |
| 15 | 16 | 17 | 18 | 19 | 20 | 21 |
| 22 | 23 | 24 | 25 | 26 | 27 | 28 |
| 29 | 30 | 31 | | | | |
| | | | | | | |

**FOCUS CHART**

*What is the weather today?*

*Today the weather is rainy.*

- **Sunny** *weather is when it's bright and clear with few or no clouds.*
- **Partly cloudy** *weather is when it's sunny but there are lots of clouds in the sky.*
- **Overcast** *weather is when the sky is gray and cloudy but it's not raining or snowing.*
- **Rainy** *weather is when it's cloudy and raining or drizzling outdoors.*
- **Snowy** *weather is when it's cloudy, and snow is falling.*

Write the weather condition words on the word wall and include the appropriate picture next to the word. Review the vocabulary daily with the weather monitor.

5. **Introduce the class calendar**
Show students the class **calendar**. Explain that the class will use it to keep track of the different kinds of weather every day for a month or two. Write the month and year at the top, and, if you have not yet done so, use a watercolor pen to number the days on the calendar.

6. **Explain the monitoring schedule**
Tell students that each of them will be the weather **monitor** for a day during the next few weeks. Each day, one weather monitor will record the day's weather. The weather monitor will stick the appropriate picture on the calendar to show each day's weather.

7. **Focus question: What is the weather today?**
Write the focus question on the chart as you read it aloud.

➤ *What is the weather today?*

When students agree on what kind of weather it is, choose the appropriate picture. Ask a student to show you today's date on the calendar. Demonstrate how to tape the weather picture to the appropriate date. (You should still be able to see the date.)

8. **Answer the focus question**
Tell students you have a strip of paper with the focus question for them to glue into their notebooks. Have students use drawings and words to answer the focus question. Be sure that students include today's date on the page.

You might model a notebook entry describing today's weather. For beginning writers, provide a sentence frame such as: Today the weather is _____ .

About once every 2 weeks, or when there is a change in the weather, have students make another weather entry in their science notebooks.

# WRAP-UP/WARM-UP

## 9. Share notebook entries

Conclude Part 1 or start Part 2 by having students share their notebook entries. Ask students to open their science notebooks to the most recent entry. Read the focus question together.

➤ *What is the weather today?*

Ask students to pair up with a partner to

- share their answers to the focus question;
- explain their drawings.

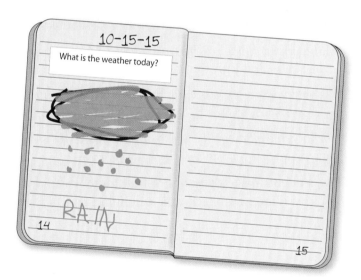

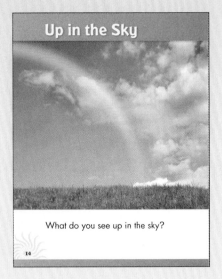

**Up in the Sky**

What do you see up in the sky?

14

## READING *in Science Resources*

**10. Read "Up in the Sky"**

Introduce the title of the article, "Up in the Sky." Ask students to brainstorm a few ideas of what they might see in the sky.

Explain that this article will tell them about the things they might see in the sky during the day and at night. Ask them to listen for ideas to add to their brainstorming list.

Read the article aloud. Pause to discuss the questions posed in the reading.

**11. Discuss the reading**

After the reading, ask,

➤ *What might you see in the sky during the day?* [Sun, clouds, birds, butterflies, airplanes, balloons, kites, parachutes.]

➤ *What might you see in the sky at night?* [Moon, stars, lights from airplanes, clouds in the moonlight.]

➤ *What things might you see in the sky at night and during the day?* [Moon, stars, clouds.]

➤ *What is the difference between day and night?* [It is light during the day from the Sun and dark at night.]

Have students make drawings in their notebooks of a day sky and a night sky. Have them share their drawings with a partner.

# MATERIALS *for*

## Part 2: *Recording Temperature*

### For each student

1   *Thermometer Outline* (See Step 6 of Getting Ready.)

### For the class

1   Garden thermometer

1   Demonstration thermometer

1   Permanent marking pen, black ★

1   Piece of fadeless construction paper of each color: orange, yellow, green, blue, purple ★

2   Basins ★

•   Warm water ★

•   Cold water ★

•   Masking tape ★

1   Ruler ★

❏   1   Teacher master 24, *Thermometer Outline*

### For assessment

•   *Assessment Checklist*

★ Supplied by the teacher.      ❏ Use the duplication master to make copies.

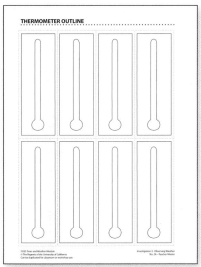

*No. 24—Teacher Master*

## GETTING READY *for*

### Part 2: *Recording Temperature*

1. **Schedule the investigation**

   This is a whole-class activity. Plan on 15 minutes to introduce the thermometer, 15 minutes to observe air temperature outdoors, and 15 minutes indoors for recording on the calendar and in notebooks.

2. **Preview Part 2**

   Students use a thermometer and take turns measuring and recording the relative temperature (freezing, cold, cool, warm, hot). The focus question is **How can we measure the air temperature?**

3. **Plan to introduce temperature**

   You will need a container of cold water and a container of warm water for students to feel when you introduce the thermometer.

4. **Consider the thermometers**

   The kit includes a large garden thermometer and a simulated thermometer for demonstration. They have both Celsius and Fahrenheit scales. Decide which scale you want to use and cover the other scale on the garden thermometer with paper. The demonstration thermometer has the Celsius scale on one side and Fahrenheit on the other.

   If it is practical, plan to mount the garden thermometer outdoors where it can be viewed from indoors. Against the window would be ideal. Otherwise, plan to send the weather-monitor team (the student monitor for the day and an adult) outdoors with the thermometer each day. It takes about 2 minutes for the thermometer to reach outdoor air temperature.

   Adjust the ribbon on the demonstration thermometer to display the day's temperature. Students note the temperature from the demonstration thermometer and record the word that describes the relative temperature (cold, cool, warm, etc.).

5. **Prepare the demonstration thermometer**

   The first time the demonstration thermometer is used, five temperature ranges need to be color coded. Students will use the color-coded areas on the thermometer to help them associate temperatures with how the air feels.

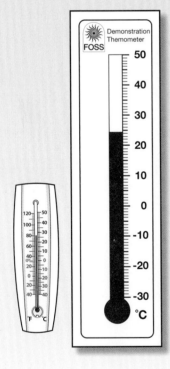

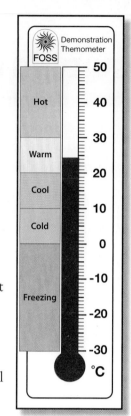

Use fadeless art paper or construction paper to code the temperature ranges listed below. Write the words in black permanent marker. Be careful not to impede the movement of the red-and-white ribbon used to change the temperature reading.

Code 30°C–50°C (80°F–120°F) orange. Label it "hot."

Code 20°C–30°C (65°F–80°F) yellow. Label it "warm."

Code 10°C–20°C (50°F–65°F) green. Label it "cool."

Code 0°C–10°C (32°F–50°F) blue. Label it "cold."

Code the area below 0°C (32°F) purple. Label it "freezing."

6. **Prepare thermometer outlines**

   Make enough copies of teacher master 24, *Thermometer Outline*, so that each student can glue one outline into their notebook when they answer the focus question in Step 11. There are eight copies of the thermometer outline on the master.

7. **Plan assessment**

   Here are specific objectives to observe in this part.

   - *Temperature is how hot or cold it is.*

   - *Thermometers measure temperature.*

   - *Air temperature tells something about the weather.*

   Focus on a few students each session. Record the date and a + or − on the *Assessment Checklist*.

## FOCUS QUESTION

**How can we measure the air temperature?**

### TEACHING NOTE

*You might compare this thermometer to the ones for measuring body temperature, which help determine if someone is sick or not.*

### EL NOTE

**Record the bold word on the word wall with an illustration.**

**Materials for Steps 2–4**
- *Garden thermometer*
- *Demonstration thermometer*
- *Warm and cold water*
- *Paper towels*
- *Basins*

## GUIDING *the Investigation*
### Part 2: *Recording Temperature*

1.  **Introduce** *temperature*
    Call students to the rug. Ask them to close their eyes and think about how the air felt when they were last outdoors. Ask,

    ➤ *How does the air outdoors feel today? Is it **hot** or **cold**?*

    ➤ *What makes the outdoors air warm or hot?* [The sun.]

    Explain,

    *Even though we can't see air, we can feel how hot or cold it is.* **Temperature** *describes how hot or cold the air is.*

    ➤ *What are some words we might use to describe the temperature?*

2.  **Introduce the thermometer**
    Hold up the garden thermometer. Say,

    *This is a _____ .*

    (Wait to see if students know the word.)

    *Everybody say "**thermometer**."*

    *A thermometer is a tool to measure temperature. We can use this thermometer to measure the temperature of the air.*

    Demonstrate how to read the air temperature of the room.

3.  **Focus on the red line**
    Show students the red line in the thermometer stem. Tell them,

    *When the thermometer gets hot, the red line moves up higher. When the thermometer gets cold, the red line goes down lower.*

    Use the demonstration thermometer to show the change of the level of the red line when the thermometer gets hot and cold.

4.  **Measure the temperature of water**
    Bring out the two containers of water. Invite students to feel the water and tell which is warm and which is cold. Ask,

    ➤ *What do you think will happen to the red line on the thermometer if we put it in the cold water? Will the red line go up or down?*

    Adjust the demonstration thermometer to represent the temperature on the garden thermometer. Point out to students,

    *This is where the red line is on our thermometer. The top of the red line is right here at <number> degrees.*

Put the thermometer in the cold water. Have students watch the red line. After a minute, read the thermometer.

*Before we put the thermometer in the cold water, the red line was here at <number> degrees.*

(Adjust the demonstration thermometer to the cold-water temperature.)

*Now the red line is here at <number> degrees.*

➤ *Did the red line go up or down?* [Down.]

➤ *Is the water colder or warmer than the room temperature?* [Colder.]

Move the thermometer to the warm water, and repeat the process. Dry off the thermometer, and allow it to sit a few minutes to register room temperature.

5. **Go outdoors**
   Before going outdoors, ask students to predict whether it is colder or warmer outdoors, compared to the classroom. Ask them what they think will happen to the red line when you take the thermometer outdoors. On the thermometer, mark the room temperature with masking tape. Then, take the class and the two thermometers to an outdoor area.

6. **Monitor the outdoor temperature**
   Once outdoors, have students sit in a circle as the thermometer responds to the outdoor air temperature. Have a student point to the top of the red line on the thermometer.

   Adjust the demonstration thermometer to represent the temperature on the garden thermometer. Ask,

   ➤ *Is the red line higher or lower than it was in our classroom?*

   ➤ *Is it warmer or cooler outdoors today?*

   Have students predict what will happen to the red line when they go back into the classroom.

7. **Return to class**
   Return to the classroom. When students are gathered at the rug, point out the five colored-coded zones along the side of the thermometer. Tell students,

   *When the red line on the thermometer is down in this purple area, the temperature is **freezing** cold. When the red line is in this blue area, the temperature is cold. The green area is **cool** weather, the yellow area is **warm**, and when the red line is in the orange area, the weather is hot.*

**Materials for Step 5**
• *Masking tape*

**TEACHING NOTE**

*Position the thermometer where direct sunshine will not fall on it.*

Say it
Write it
**New Word**
See it
Hear it

| October | | | | | | |
|---------|--------|---------|-----------|----------|--------|----------|
| Sunday | Monday | Tuesday | Wednesday | Thursday | Friday | Saturday |
| 1 ☀ | 2 ☀ | 3 ☀ | 4 ☁ | 5 ☁ | 6 ☁ | 7 ☀ |
| 8 ☀ | 9 ☁ | 10 ☀ warm | 11 | 12 | 13 | 14 |
| 15 | 16 | 17 | 18 | 19 | 20 | 21 |
| 22 | 23 | 24 | 25 | 26 | 27 | 28 |
| 29 | 30 | 31 | | | | |

cold
cool
freezing
hot
temperature
thermometer
warm

## EL NOTE

**Provide beginning writers with a sentence frame such as:
The air is _____ .**

## FOCUS CHART

*How can we measure the air temperature?*

*We use a thermometer to tell how hot or cold the air is.*

8. ### Discuss the weather
   Ask the students to think back to their outdoor experience. Ask,

   ➤ *Was the weather warm, cool, or cold outdoors today?*

9. ### Explain temperature monitoring
   The class weather monitor (working with an adult) will measure the outdoor temperature each day and help set the demonstration thermometer for the class. The word describing the air temperature will be recorded on the calendar by an adult. The monitor will tape the picture for the day's weather to the calendar.

   Write the word that represents the temperature on the calendar.

10. ### Review vocabulary
    This is a good time to review the vocabulary that students can use to describe the air temperature. They should use this vocabulary when they record the weather on the calendar and in their science notebooks.

11. ### Focus question: How can we measure the air temperature?
    Write the focus question on the chart as you read it aloud.

    ➤ *How can we measure the air temperature?*

    Distribute copies of the focus question and teacher master 24, *Thermometer Outline*, for students to glue into their notebooks. Have students fill in the thermometer outline and use words to answer the focus question.

# WRAP-UP/WARM-UP

12. ### Share notebook entries
    Conclude Part 2 or start Part 3 by having students share notebook entries. Ask students to open their science notebooks to the most recent entry.

    ➤ *How can we measure the air temperature?*

    Ask students to pair up with a partner to

    - share their answers to the focus question;

    - explain their drawings.

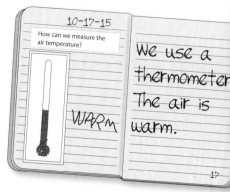

# MATERIALS *for*

## Part 3: *Wind Direction*

### For each student

- 1   Piece of construction paper,  10 × 23 cm (4" × 9") ★
- 8   Strips of crepe paper, 2 × 25 cm (1" × 10")
      (See Step 4 of Getting Ready.)
- 2   Pieces of yellow yarn, 60 cm (24")
- 1   Piece of transparent tape, 9 cm (3.5") ★
- 1   *FOSS Science Resources:  Trees and Weather*
    - • "Weather"

### For the class

- 1   Hole punch
- •   Glue stick or white glue ★
- 1   Scissors ★
- ❏ 1   Teacher master 25, *Center Instructions—Wind Direction*
- 1   Big book, *FOSS Science Resources:  Trees and Weather*

### For assessment

- •   *Assessment Checklist*

★ Supplied by the teacher.          ❏ Use the duplication master to make copies.

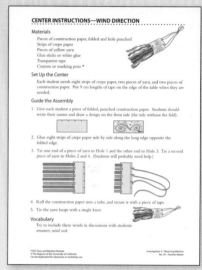

*No. 25—Teacher Master*

# GETTING READY *for*

## Part 3: *Wind Direction*

1. **Schedule the investigation**
   This part can be done as a whole-class activity or at a center. Plan 15–20 minutes for an introduction, 15–20 minutes for wind-sock assembly, and 10 minutes to fly the wind socks outdoors. Students will need 10 minutes for notebook entries and 15 minutes for the reading.

2. **Preview Part 3**
   Students construct a wind sock and observe how it responds when air moves through it. They find out that they can determine wind direction by using a wind sock. The focus question is **What does a wind sock tell us about the wind?**

**▶NOTE**
You can use thicker paper such as card stock or tagboard. With thicker paper, you don't need to fold the edge.

3. **Prepare construction paper**
   Cut a piece of construction paper for each student. A single sheet of 23 × 30 cm (9" × 12") construction paper yields three wind socks.

   On each sheet, fold over about 1.5 cm on one long edge, and make a crease. This folded edge will be strong enough to hold the yarn handles.

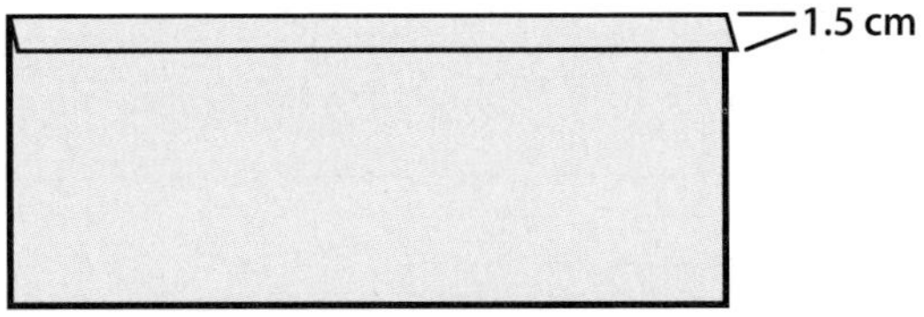


   Punch four holes through the folded edge. Space the holes out somewhat evenly: 3 cm, 8.5 cm, 14 cm, and 19.5 cm from the end.

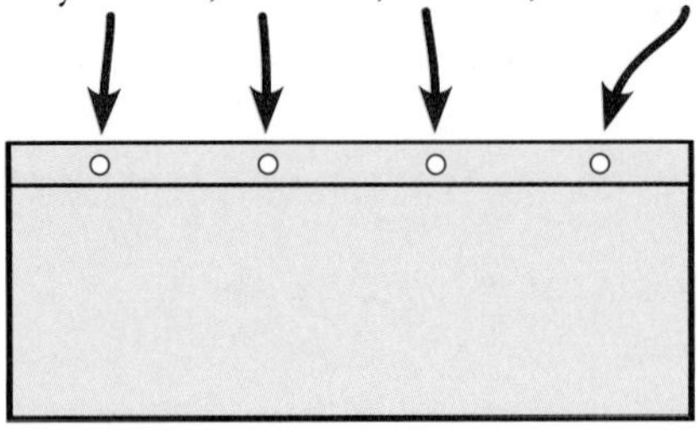

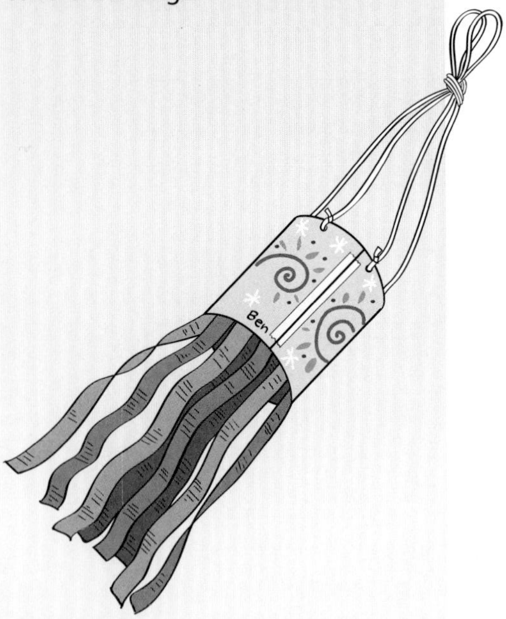


**▶NOTE**
An alternative to punching holes and tying string is to use tape to attach the string to the paper.

4. **Cut crepe-paper strips**
   Each student needs eight strips of crepe paper, 2 × 25 cm. From the roll of crepe paper, cut four 25 cm lengths per student. Cut each in half lengthwise to get eight strips.

5. **Cut yarn**

Each student will need two pieces of yarn about 60 cm long for their wind sock. One easy way to cut these is to find a piece of cardboard or a thin book that measures 30 cm in one direction. Wrap yarn around the book or cardboard two times for each student and then cut through the string bundle at one end of the cardboard. Repeat this process until you have enough yarn for students. Cut 2–4 extra pieces of yarn for repairs.

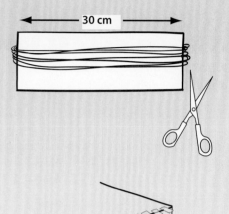

6. **Plan for tape distribution**

Each student will need a piece of transparent tape 9 cm (3.5") long to secure the wind sock paper into a tube. Put a number of pieces of tape along the edge of a table or counter for students to get quickly during the activity. Or you can provide each student with the appropriate length of tape as needed.

7. **Make a model wind sock**

Make a model wind sock using the instructions in Step 2 of Guiding the Investigation. Take the wind sock outdoors in a breeze to experience what students will observe.

8. **Select your outdoor site**

Walk around your schoolyard and select a good location on the schoolyard for students to use their wind socks. The area should be exposed to the wind and be large enough for students to spread out and move freely with their wind socks.

9. **Check the site**

It is always a good idea to check the outdoor site on the morning of an outdoor activity. Check for any distracting items or unsafe items where students will be working.

10. **Plan to read** *Science Resources*: **"Weather"**

Plan to read "Weather" during a reading period near the end of this part.

11. **Plan assessment**

Here are specific objectives for this part.

- *Wind is moving air.*

- *A wind sock indicates wind direction.*

Focus on a few students each session. Record the date and a + or − on the *Assessment Checklist*.

## FOCUS QUESTION

**What does a wind sock tell us about the wind?**

Say it

**New Word**

See it

Hear it

Write it

### Materials for Step 2

- *Construction paper, with holes*
- *Crepe-paper strips*
- *Yarn*
- *Transparent tape*
- *Glue stick or white glue*

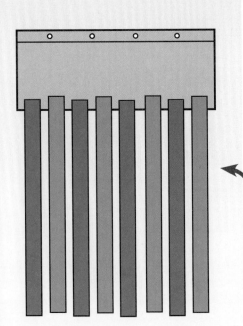

# GUIDING *the Investigation*

## Part 3: *Wind Direction*

1. **Introduce wind socks**

   Call students to the rug. Show them your wind sock and ask,

   ➤ *This is a* **wind sock**. *What do you think it can be used for?*

   Tell students that **wind** is **moving air**. Discuss how wind socks are used at airports to tell people how hard the wind is **blowing** and in what **direction**. Tell students that they are going to make wind socks, and they will go outdoors to find out which way the wind is blowing.

2. **Guide wind-sock construction**

   Send students to a learning center or prepare to make wind socks with the whole class. Distribute a piece of folded, hole-punched construction paper to each student. Have students complete each step as you demonstrate. Students will probably need help tying yarn (Step d).

   a. *Lay your piece of paper on the table with the folded edge down. Write your name on this side.*

   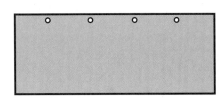

   b. *Draw a design on this side of the construction paper, using crayons or marking pens.*

   c. *Lay the construction paper on the table with the design side down. Using a glue stick, glue eight strips of crepe paper along the edge of the construction paper without the holes. The crepe paper will be the* **streamers** *that will catch the wind and show the wind direction.*

d. Tie one end of a piece of yarn to the first hole. Tie the other end to the third hole. Tie the ends of a second piece of yarn to the other two holes.

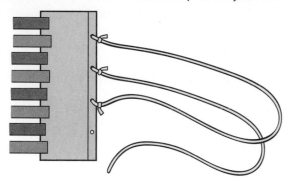

e. Roll the construction paper into a tube and tape it.

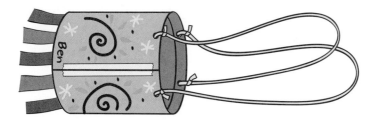

f. Gather the yarn loops and tie the ends into a knot to form a handle.

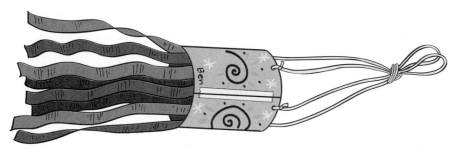

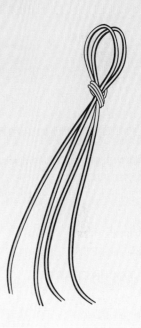

### 3. Go outdoors

When the wind socks are complete, go outdoors to test them. If there is no wind, wait to test them when students can observe how their wind socks interact with the wind. The streamers should catch the wind and point in the direction the wind is blowing. Have students stand still and hold the wind sock out in front of them. Ask,

➤ *What happens to your wind sock?* [The sock catches the wind and the streamers fly out to the side.]

➤ *How can you tell the direction the wind is blowing?* [The top of the wind sock faces into the wind. The streamers in the tail fly out in the direction the wind is blowing.]

**4. Return to class**

Back in the classroom, add a description of today's wind to the calendar. Work with the students to come up with the description such as no wind (calm), little wind (breezy), lots of wind (windy).

Secure a wind sock in a location where air moves (open window, open door, air duct) so students can observe the air movement.

You can also make a wind sock out of more durable materials to hang outdoors where it can be viewed from the window or at recess.

**5. Review vocabulary**

Add to the word wall any words that came up in the outdoor activity.

**6. Focus question: What does a wind sock tell us about the wind?**

Now that students have experienced using the wind sock, ask the focus question.

➤ *What does a wind sock tell us about the wind?*

Write the focus question on the chart as you read it aloud. Distribute the focus-question strips for students to glue into their notebooks. Have students use pictures and words to answer the focus question.

---

*blowing*
*direction*
*moving air*
*streamer*
*wind*
*wind sock*

---

**EL NOTE**

**Provide beginning writers with a sentence frame such as:**
**I noticed _____ .**

---

**FOCUS CHART**

What does a wind sock tell us about the wind?

Wind socks show the direction of the wind.

---

# WRAP-UP

### 7. Share notebook entries

Conclude Part 3 by having students share notebook entries. Ask students to open their science notebooks to the most recent entry. Read the focus question together.

➤ *What does a wind sock tell us about the wind?*

Ask students to pair up with a partner to

- share their answers to the focus question;
- explain their drawings.

**TEACHING NOTE**

*See the **Home/School Connection** for Investigation 3 at the end of the Interdisciplinary Extensions section. This is a good time to send it home with students.*

**Weather**

Weather is in the air.
Sometimes the weather is hot.
Sometimes the weather is cold.

24

## READING *in Science Resources*

8.  **Read "Weather"**

    Introduce the title, "Weather." Ask students for words that describe weather.

    Tell students that this article describes different weather conditions. Ask them to see if the weather conditions described in the article are ones they have experienced or observed.

    Read the article aloud. Pause to discuss the questions posed in the article.

9.  **Discuss the reading**

    After the reading, ask,

    ➤ *What are the changes that we see when the weather changes?* [We see clouds change, sunshine come or go, rain fall, things blowing in wind.]

    ➤ *What are the changes that we feel when the weather changes?* [The temperature gets hotter or colder, everything gets wet if it rains, air blows things around when it gets windy.]

    ➤ *Where is weather?* [In the air.]

    ➤ *What tools do we use to measure weather?* [Thermometer, rain gauge, and wind sock.]

# INTERDISCIPLINARY EXTENSIONS

## Language Extension

- **Read weather literature and poetry**

  Many children's books and collections of poetry by both children and adults complement these activities. *The Wind*, by Monique Felix, is a picture essay without words that children will enjoy.

## Math Extension

- **Make a temperature bar graph**

  Make a daily temperature graph using chart paper with 1" squares. Cut strips of red paper 2 × 30 cm (3/4" × 12"). As the class weather monitor reports the daily temperature on the demonstration thermometer, cut a strip of paper to match the height of the red line on the thermometer. Glue each strip on the graph paper to make a graph of daily changes.

## Art Extensions

- **Create a wind catcher**

  Have students design and construct wind catchers. Provide a variety of craft materials at a center: straws, crepe-paper scraps, paper, cardboard, plastic bags, string, yarn, thread, pipe cleaners, paper cups, paper plates, feathers, fabric remnants, and tape or glue. After students have completed their projects, ask,

  ➤ *Does it move? What makes it move?*

  ➤ *Why do you think this is a wind catcher?*

  ➤ *What does it tell you about the wind?*

- **Bring wind catchers from home**

  Have students bring in wind catchers they might have at home to share with the class. These might include wind chimes, mobiles, and wind socks.

> **TEACHING NOTE**
>
> *Refer to the teacher resources on FOSSweb for a list of appropriate trade books that relate to this module.*

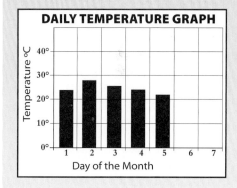

- ## Make carp wind socks

  *Koinobori* are carp-shaped streamers, or carp wind socks, used as decorations to celebrate Children's Day (originally Boys' Festival) on May 5 in Japan. In Japanese culture, the carp symbolizes courage and strength because of its ability to swim upstream against a strong current. Use the Internet to search for patterns and information on how to make carp wind socks.

## Home/School Connection

Students make wind chimes using paper cups as bases and a variety of metal objects that make a sound when they hit each other. Large nails make effective wind chimes, but care must be taken when working with sharp objects. Bolts can be substituted for nails.

Make copies of teacher master 26, *Home/School Connection* for Investigation 3, and send it home with students at the end of Part 3.

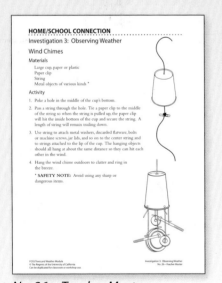

*No. 26—Teacher Master*

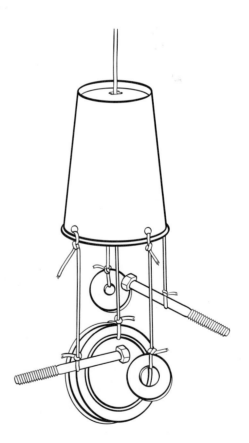

**Part 1**
Fall: What Comes from
Trees? .................................. **170**

**Part 2**
Fall: Food from Trees ............ **173**

**Part 3**
Fall: Visiting Adopted Trees ... **177**

**Part 4**
Winter: Evergreen Hunt ....... **182**

**Part 5**
Winter: Twigs ...................... **187**

**Part 6**
Winter: Visiting Adopted
Trees ................................... **191**

**Part 7**
Spring: Forcing Twigs ........... **195**

**Part 8**
Spring: Bark Hunt .............. **198**

**Part 9**
Spring: Visiting Adopted
Trees ................................... **202**

# PURPOSE

## Content

- Seasons change in a predictable annual pattern: fall, winter, spring, and summer.

- Trees are living, growing plants.

- Bark, twigs, leaves, buds, flowers, fruits, and seeds are parts of trees.

- The buds on twigs grow into leaves or flowers.

- Some trees lose their leaves in winter; others do not.

- Trees change through the seasons.

- Some trees produce seeds that can grow into new trees of the same kind.

## Scientific and Engineering Practices

- Observe and describe seasonal changes in trees.

- Describe weather changes from season to season.

- Communicate observations and comparisons of trees.

| | Investigation Summary | Time | Focus Question |
|---|---|---|---|
| **PART 1** | **Fall:  What Comes from Trees?** Students visit the schoolyard to look for objects from trees.  Indoors, they make a chart of what they collected. | **Outdoors** 10–15 minutes **Center** 10–15 minutes | What do fall trees look like? |
| **PART 2** | **Fall:  Food from Trees** Students search for, observe, and compare seeds found in the fruits that come from trees. | **Center** 20–30 minutes **Reading** 15 minutes | What do fall trees look like? |
| **PART 3** | **Fall:  Visiting Adopted Trees** Students visit their adopted schoolyard trees. They observe the trees' bark, twigs, leaves, flowers, fruit, and seeds and add information to the class scrapbook. | **Outdoors** 15 minutes **Notebooks** 10 minutes | What do fall trees look like? |
| **PART 4** | **Winter:  Evergreen Hunt** Students hunt for evergreen trees that match samples of needles from schoolyard trees. | **Introduction** 5–10 minutes **Outdoors** 20 minutes **Reading** 15 minutes | What do winter trees look like? |
| **PART 5** | **Winter:  Twigs** Students observe the inside of tree twigs and look for growth rings, buds, and leaf scars. | **Center** 20 minutes **Reading** 15 minutes | What do winter trees look like? |

# At a Glance

| Content | Writing/Reading | Assessment |
|---|---|---|
| • Trees are living, growing plants.<br>• Bark, twigs, leaves, buds, flowers, fruits, and seeds are parts of trees. | **Science Notebook Entry**<br>Draw or write words to answer the focus question (optional). | **Embedded Assessment**<br>Teacher observation |
| • Bark, twigs, leaves, buds, flowers, fruits, and seeds are parts of trees.<br>• Seeds grow into the same kind of plant as the parent tree. | **Science Notebook Entry**<br>Draw or write words to answer the focus question (optional).<br>**Science Resources Book**<br>"My Apple Tree" | **Embedded Assessment**<br>Teacher observation |
| • Trees are living, growing plants.<br>• Trees change through the seasons. | **Science Notebook Entry**<br>Draw or write words to answer the focus question. | **Embedded Assessment**<br>Teacher observation |
| • Bark, twigs, leaves, buds, flowers, fruits, and seeds are parts of trees.<br>• Some trees lose their leaves in winter; others do not. | **Science Notebook Entry**<br>Draw or write the answer to the focus question (optional).<br>**Science Resources Book**<br>"Orange Trees" | **Embedded Assessment**<br>Teacher observation |
| • Bark, twigs, leaves, buds, flowers, fruits, and seeds are parts of trees.<br>• Twigs have structures such as leaf scars and buds. | **Science Notebook Entry**<br>Draw or write words to answer the focus question (optional).<br>**Book**<br>*Our Very Own Tree* | **Embedded Assessment**<br>Teacher observation |

| | Investigation Summary | Time | Focus Question |
|---|---|---|---|
| **PART 6** | **Winter: Visiting Adopted Trees**<br>Students revisit their adopted trees to observe any changes to the twigs, leaves, and areas around their trees. | **Introduction**<br>10 minutes<br>**Outdoors**<br>15 minutes<br>**Notebook**<br>10 minutes | **What do winter trees look like?** |
| **PART 7** | **Spring: Forcing Twigs**<br>Students bring twigs into the warmth of the classroom to force them to bloom or put out leaves. | **Center**<br>10–15 minutes | **What do spring trees look like?** |
| **PART 8** | **Spring: Bark Hunt**<br>Students observe and compare bark on a variety of trees as they search for matches to photos of the bark on schoolyard trees. | **Introduction**<br>5–10 minutes<br>**Outdoors**<br>20–25 minutes<br>**Scrapbook**<br>10 minutes | **What do spring trees look like?** |
| **PART 9** | **Spring: Visiting Adopted Trees**<br>Students revisit their trees. They look for evidence of new growth in the leaves and flowers. | **Introduction**<br>5–10 minutes<br>**Outdoors**<br>15 minutes<br>**Notebook**<br>10 minutes<br>**Reading**<br>10 minutes | **What do spring trees look like?** |

| Content | Writing/Reading | Assessment |
|---|---|---|
| • Trees are living, growing plants.<br>• Bark, twigs, leaves, buds, flowers, fruits, and seeds are parts of trees.<br>• Trees change through the seasons. | **Science Notebook Entry**<br>Draw or write words to answer the focus question. | **Embedded Assessment**<br>Teacher observation |
| • Bark, twigs, leaves, buds, flowers, fruits, and seeds are parts of trees.<br>• The buds on twigs grow into leaves or flowers. | **Science Notebook Entry**<br>Draw or write words to answer the focus question (optional). | **Embedded Assessment**<br>Teacher observation |
| • Bark, twigs, leaves, buds, flowers, fruits, and seeds are parts of trees.<br>• Trees can be identified by the pattern of the bark. | **Science Notebook Entry**<br>Draw or write words to answer the focus question (optional). | **Embedded Assessment**<br>Teacher observation |
| • Trees are living, growing plants.<br>• Trees change through the seasons.<br>• Seasons change in a predictable annual pattern: fall, winter, spring, and summer. | **Science Notebook Entry**<br>Draw or write words to answer the focus question.<br>**Science Resources Book**<br>"Maple Trees" | **Embedded Assessment**<br>Teacher observation |

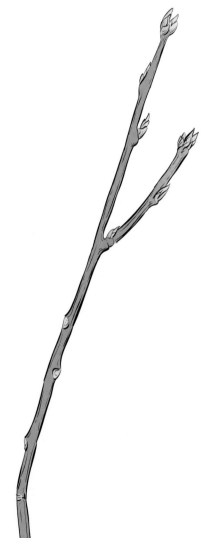

# BACKGROUND *for the Teacher*

Trees don't move from place to place, respond to stimuli instantaneously, make sounds, or place obvious demands on their environment, so it is easy for them to fade into the scenery and be overlooked as important living organisms. Often, it takes months or even years to observe evidence that trees are alive and that important things are going on that ensure their survival and reproduction. Trees do change, and sometimes it is necessary to observe trees through the seasons to become aware of their changes.

One change is growth. In most cases, tree growth is slow. Trees grow in two ways. Their structures (trunk, branches, and twigs) get thicker by adding layers of woody material to their circumference (like dipping a candle in melted wax to increase its size), and they get longer by extending new twigs from the ends and sides of old ones.

Two different tissues are responsible for these two kinds of growth. Tissue called cambium adds bulk to the tree. The cambium layer covers the entire tree like a glove covers a hand. This vital part of the tree adds wood cells on its inner side (like the inside of the glove adding a new layer of skin to your finger), and generates bark on its outer layer. When a trunk or branch is crosscut, this layer can be seen, as well as the concentric layers of wood laid down over previous years. The age of a twig, branch, or trunk can often be determined by counting the **growth rings** of wood that are visible.

Another kind of plant tissue results in the lengthening of tree structures. This growth takes place at **buds**. Buds can be found at the tip of last year's twigs, or along the sides of more mature twigs and branches. When the bud receives the signal to snap into action (usually the warming of spring), the cells begin to lengthen and multiply, making the old structure longer.

A growing tip is the only place that a tree gets longer. Trees do not grow taller by growing out of the ground. If you had a favorite childhood swing suspended from a branch of a tree in your yard, that branch would be exactly the same height above the ground today.

## What Do Fall Trees Look Like?

Over time, trees have evolved different strategies for coping with harsh **winter** temperatures. Some trees simply rid themselves of their sensitive leaves when winter approaches. Trees that drop their leaves are deciduous. For many trees, the transition between the productive **summer season** and the dormant winter season includes a brief, often extravagant, display of color. The **fall** colors are revealed when the dominant green pigment (chlorophyll) decomposes, allowing the red and yellow pigments to surface. It is during this time that the last of the ripening **fruits** and seeds fall from the trees, leaving the deciduous trees bare and dormant.

## What Do Winter Trees Look Like?

After the brassy fall foliage falls to the ground and the energy-laden sap retreats to the roots, most of the metabolic functions of deciduous trees shut down for the winter months. If you inspect the branches and twigs during this dormancy, **leaf scars** show where each of last year's leaves was attached, and often darkened rings around a twig show the extent of the last several seasons' growth. The buds that hold the promise of next year's growth can be clearly seen.

Other trees, the **evergreens**, hold their leaves throughout the year. The wintry image of a conifer—pine, spruce, or fir—cloaked in snow is a familiar one. Conifers have evolved so that the leaves (either **needles** or **scales**) are much reduced in size, and the sap is thick, sticky, and freeze-resistant, allowing these trees to continue their metabolic processes throughout the winter. Broadleaf evergreens that live in temperate climates rarely, if ever, experience freezing temperatures, so they can keep their leaves for years. If these trees, such as citrus and magnolia, do experience a hard freeze, they may be seriously damaged or even die.

## What Do Spring Trees Look Like?

Most trees acknowledge the coming of spring with a resurgence of sap up the trunks and branches. The injection of energy stimulates the buds to start developing. Three kinds of buds can be located on the twigs of typical deciduous trees: leaf, stem, and reproductive structures. Often, the first to answer the call of longer, warmer days are the reproductive buds. The famously popular pussy willow buds of the willow tree swell and transform into comely little fuzzy flowers. **Blossoms** and catkins spring out on trees in a wave that starts in the south and surges north through March, April, and May.

As blossoms fade, fruit develop, fulfilling the promise of species perpetuation. Simultaneously, leaf buds produce the profusion of leaf biomass that constitutes the **food** factory for the tree. Other buds extend new twigs and branches to provide additional structure for increasing the reach of the energy trap.

The spring tree is lush and vibrant. The tree is never greener, and production is at its highest. This is when the greatest amount of growth can be observed—new leaves, new branches, new fruits and seeds. The tree is busy laying down a ring of wood just under the bark. You might be able to measure the girth at the end of spring and confirm that your favorite tree is more massive than it was at the start of fall.

**New Word**

Say it · See it · Hear it · Write it

*Blossom*
*Bud*
*Evergreen*
*Fall*
*Food*
*Forcing*
*Fruit*
*Growth ring*
*Leaf scar*
*Needle*
*Scale*
*Season*
*Spring*
*Summer*
*Swollen*
*Winter*

# TEACHING CHILDREN *about*
## *Trees through the Seasons*

Early-childhood students may overlook the fact that trees are alive and going about their business in a slow but predictable manner. This investigation provides a few strategies for helping young students discover some of the indicators of life and start appreciating the large rhythms that guide the lives of our arboreal neighbors.

The investigations are organized into seasonal activities: three fall parts, three winter parts, and three spring parts. The **conceptual flow** starts with students reviewing that **trees are living plants**. In Parts 1–3, during an excursion to the schoolyard, students find that some **tree parts are found on the ground** under trees: seeds, twigs, leaves, and fruit. Students investigate the kinds of **food people get from trees**: fruits and seeds. During a last look at fall trees, students remove a leaf from a twig to observe the **leaf scar** where the leaf attached to the twig.

In Parts 4–6, students visit trees during the winter. They observe that some trees, called **evergreen trees, stay green all winter**. The **conifers have specialized leaves** called **needles** and **scales**. Students look closely at **structures on dormant twigs** of **deciduous** trees. They observe **growth rings**, **leaf scars**, and **buds**.

In Parts 7–9, students continue investigating tree twigs as spring approaches. They place cut twigs in water and observe that buds produce **flowers** and **leaves**. Students use reference images of tree bark to identify trees by **bark pattern**. After observing their adopted trees throughout the year, students start to understand that **trees change through the seasons**.

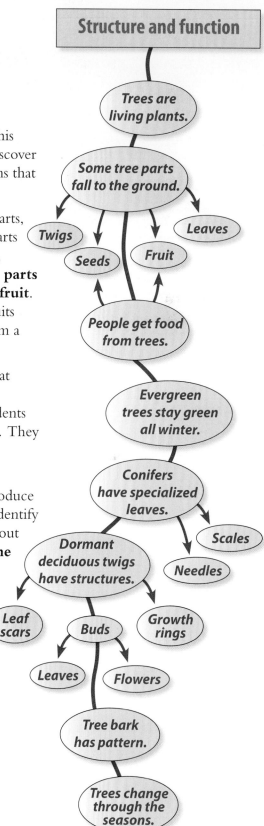

**FOCUS QUESTIONS B**

Inv. 3, Part 1:  What is the weather today?

Inv. 3, Part 2:  How can we measure the air temperature?

Inv. 3, Part 3:  What does a wind sock tell us about the wind?

Inv. 4, Part 1:  What do fall trees look like?

Inv. 4, Part 2:  What do fall trees look like?

Inv. 4, Part 3:  What do fall trees look like?

Inv. 4, Part 4:  What do winter trees look like?

Inv. 4, Part 5:  What do winter trees look like?

Inv. 4, Part 6:  What do winter trees look like?

Inv. 4, Part 7:  What do spring trees look like?

Inv. 4, Part 8:  What do spring trees look like?

Inv. 4, Part 9:  What do spring trees look like?

FOSS Trees and Weather Module
© The Regents of the University of California
Can be duplicated for classroom or workshop use

Investigations 3–4
No. 3—Teacher Master

*No. 3—Teacher Master*

# MATERIALS *for*

## Part 1: *Fall: What Comes from Trees?*

For each pair of students

1  Zip bag, 1 L

For the class

1  Plastic bag, 1 L ★

•  Chart paper ★

•  Transparent tape (optional) ★

❏  1  Teacher master 3, *Focus Questions B*

For assessment

•  *Assessment Checklist*

★ Supplied by the teacher.          ❏ Use the duplication master to make copies.

# GETTING READY *for*

## Part 1: *Fall: What Comes from Trees?*

1. **Schedule the investigation**
   Plan 10–15 minutes outdoors with the whole class and 10–15 minutes for the whole class or groups of six to eight inside at a center.

2. **Preview Part 1**
   Students visit the schoolyard to look for objects from trees. Indoors, they make a chart of what they collected. The focus question is **What do fall trees look like?**

3. **Prepare the bar graph on chart paper**
   Students organize items they find on a class bar graph. Draw five columns on a large piece of chart paper. Label the bottom of the columns "Seeds," "Leaves," "Twigs," "Other," and "Not from a tree."

4. **Select your outdoor site**
   Consider the trees that you would like students to visit and how best to have students collect different materials under those trees. If possible, allow pairs of students to select and visit trees on their own. Otherwise, select several different trees to visit and lead a walk from tree to tree. Give students time at each tree to collect items in their bags.

   It is always a good idea to check the outdoor site on the morning of an outdoor activity. Check for any distracting items or unsafe items where students will be working.

5. **Plan word wall, focus charts, and notebook entries**
   Add words to the word wall informally in each part. Formal suggestions for the word wall and focus charts appear only at the end of each season's activities (Parts 3, 6, and 9).

   Each part has a focus question and the opportunity for students to respond to the question in their notebooks. Plan when to have students make entries in their notebooks.

6. **Plan assessment**
   Here are the specific objectives to observe in this part.

   - *Tree are living, growing plants.*

   - *Bark, twigs, leaves, buds, flowers, fruits, and seeds are parts of trees.*

   Focus on a few students each session. Record the date and a +or − on the *Assessment Checklist*.

| | | | | Not from a tree |
|---|---|---|---|---|
| Seeds | Leaves | Twigs | Other | |

**TEACHING NOTE**

*You could have the students respond to the focus question after each part or have the students make a drawing in their notebooks at the end of each season.*

*What do fall trees look like?*

**Materials for Step 2**
- *Zip bags*
- *Plastic bag*

*As students sort their treasures, write a few key words on the word wall.*

**Materials for Step 3**
- *Bar graph on chart paper*

# GUIDING *the Investigation*
## Part 1: *Fall: What Comes from Trees?*

1. **Introduce the investigation**

   Call students to the rug. Have them recall the kinds of things they found when they took walks to look at trees and leaves. Tell them they will be going outdoors to look for things on the ground underneath the trees. Ask,

   ➤ *What things do you think you will find under trees today?*

   ➤ *How could you find out if the things you find under a tree came from that tree?*

   ➤ *Do you think you will find anything that isn't part of a tree?*

2. **Go outdoors**

   Organize the class into pairs. Give each pair of students a zip bag. Move outdoors in the usual orderly manner.

   Form a sharing circle. Tell students to make a collection of ten things they find *under* trees. Set boundaries so students know the limits of their search area, or lead them on a walk to visit several different trees. Make your own collection of seeds for Part 2.

3. **Sort the tree items**

   If the weather is good, remain outdoors to sort the items on the bar graph. If not, return to the classroom and help students sort their collections into categories on the graph. Explain the categories: "seeds," "leaves," "twigs," "other," and "not from a tree."

   When the sorting is complete, ask,

   ➤ *What did you find the most of?*

   ➤ *How could you find out which tree these things came from?*

   ➤ *What was placed in the "Not from a tree" column? Are any of those things nonliving?*

   ➤ *Do you think you would find different things if we looked under trees in the winter? In the spring?*

4. **Clean up**

   Have students help you save any seeds that were gathered. The seeds will be used in Part 2. Save a few items from each of the other columns All other items can be returned outdoors. Keep the graph and saved items at a learning center for students to mix up and sort again, or tape the items to the graph for display.

# MATERIALS *for*

## Part 2: *Fall: Food from Trees*

### For each student

1   Piece of drawing paper, 11 × 14 cm (4.25" × 5.5") ★

1   *FOSS Science Resources: Trees and Weather*

   • "My Apple Tree"

### For the class

• Edible fruits ★

• Seeds (from Part 1) ★

1   Knife ★

• Paper towels ★

• Crayons or colored pencils ★

• Transparent tape ★

• White glue ★

1   Piece of chart paper ★

❏  1   Teacher master 27, *Center Instructions—Food from Trees*

1   Big book, *FOSS Science Resources: Trees and Weather*

### For assessment

• *Assessment Checklist*

★ Supplied by the teacher.    ❏ Use the duplication master to make copies.

*No. 27—Teacher Master*

## GETTING READY *for*
### Part 2: *Fall: Food from Trees*

1.  **Schedule the investigation**
    The review of seeds collected in Part 1 can be done as a whole class activity or at the center. Work with groups of six to ten students at the center. Plan 20–30 minutes for each group. In addition, plan 15 minutes for the reading.

2.  **Preview Part 2**
    Students search for, observe, and compare seeds found in the fruits that come from trees. The focus question is **What do fall trees look like?**

3.  **Check for allergies**
    Before beginning this part, be certain you know of any students who have allergies to any fruits. Be sure to wash all fruits that students will touch.

    Plan to discuss safety with the students and remind them not to put anything in their mouths unless the teacher tells them it is OK to do so.

    If students will be tasting selected fruits, have them wash their hands before eating the fruit.

4.  **Obtain fruits**
    Get several kinds of fruits that grow on trees. Include common fruits, such as apples, tangerines, oranges (with seeds), pears, cherries, and plums. You might ask parents to help. You'll need one item for each student. Larger fruits, such as oranges, apples, and pears, can be shared by four to six students.

5.  **Plan to read** *Science Resources*: **"My Apple Tree"**
    Plan to read "My Apple Tree" during a reading period near the end of this part.

6.  **Plan assessment**
    Here are specific objectives for this part.

    *   *Bark, twigs, leaves, buds, flowers, fruits, and seeds are parts of trees.*

    *   *Seeds grow into the same kind of plant as the parent tree.*

    Focus on a few students each session. Record the date and a + or − on the *Assessment Checklist*.

# **GUIDING** *the Investigation*
## Part 2: *Fall: Food from Trees*

1. **Review seeds**

   Distribute some of the seeds you and your students collected in Part 1. Let students explore the seeds and compare them. Ask,

   ➤ *What are seeds?* [The part of the plant from which a new plant can grow.]

   ➤ *If we planted one of the seeds from the oak tree, what kind of plant would grow?* [A new oak tree.]

   Explain that each parent plant produces plants of its own kind. The parent plant produces baby oak trees from its seeds. Remind the class about the book *Our Very Own Tree*.

   ➤ *Do the seeds all look the same? How are they different?* [Focus on color, size, shape, and texture.]

2. **Have students wash hands**

   Collect the seeds. Students will be touching and eating food at the center. Begin by having students wash their hands with warm, soapy water before working with food at the center.

3. **Distribute fruits**

   Show students at the center the fruits you brought to class. Explain that these fruits are parts of trees. They are also food that we and other animals eat. Cut the fruits. Give each student a paper towel and a piece of fruit that includes a seed. Ask students to see if they can find a seed inside the fruit.

   As students observe the seeds and fruit, write on the class word wall a few key words they use or need (*food, fruit*).

4. **Record the findings**

   Distribute more paper towels, paper, pencils, crayons, and glue or tape. Have each student draw the fruit on a piece of paper, then tape or glue a seed next to the drawing. Combine all the drawings into one class poster.

5. **Taste the fruit**

   Check your class list of allergies before allowing students to eat any fruit. Invite students to taste the different fruits. Cut pieces from the fruits that students observed and serve them to the group.

6. **Prepare for the next group**

   Recycle any food remains, saving any untouched slices of larger fruits for the next group.

**Materials for Step 1**
• *Seeds from Part 1*

**Materials for Steps 3–4**
• *Fruit*
• *Knife*
• *Paper towels*
• *Glue*
• *Tape*
• *Drawing paper*
• *Crayons or colored pencils*

▶ **SAFETY NOTE**
Remind students that they can only eat during science investigations if they are told it is OK to do so by the teacher.

# READING *in Science Resources*

7. **Read "My Apple Tree"**

   This is the first of three articles about common trees through the seasons. Students have just observed the fruit and seeds of the apple tree, and this article is about that tree. The article starts with winter, moves to spring and summer, and ends with fall.

   Read the article aloud. Pause to discuss any tree structures that are mentioned in the reading.

8. **Discuss seasons and trees**

   After the reading, ask,

   ➤ *What do we call winter, spring, summer, and fall?* [**Seasons**.]

   ➤ *What season is it now where we live?* [Fall.]

   ➤ *We have been recording the weather on our calendar. What kind of weather have we had this fall?*

   ➤ *What happens to the apple tree in fall?* [Apples are ready to eat. Leaves start falling off the apple tree.]

   ➤ *Do leaves fall off any other kinds of trees in fall?*

   ➤ *What season comes after fall?* [Winter.]

   ➤ *How will the winter weather be different from our fall weather?*

   ➤ *What does an apple tree look like during the winter?* [It is leafless.]

   ➤ *What happens to the winter ice and snow in the spring?* [It melts.]

   ➤ *What causes the ice and snow to melt?* [The Sun makes it warm and the frozen water melts.]

**My Apple Tree**

It is so cold in winter!
Water freezes into ice.
There is ice on the twigs.

36

# MATERIALS
## Part 3: *Fall: Visiting Adopted Trees*

### For the class

- 1 Weather calendar (from Investigation 3)
- • String
- 1 Garden thermometer
- 1 Demonstration thermometer
- 1 Camera (optional) ★
- 1 Scissors ★
- 1 Tree scrapbook

### For assessment

- • *Assessment Checklist*

★ Supplied by the teacher

## GETTING READY *for*

### Part 3: *Fall: Visiting Adopted Trees*

1. **Schedule the investigation**
   Plan on 15 minutes outdoors for students to observe their adopted trees and 10 minutes to draw and write in their notebooks in the classroom.

2. **Preview Part 3**
   Students visit their adopted schoolyard trees. They observe the trees' bark, twigs, leaves, flowers, fruit, and seeds and add information to the class scrapbook. The focus question is **What do fall trees look like?**

3. **Check the outdoor site**
   It is always a good idea to check the outdoor site on the morning of an outdoor activity. Check for any distracting items or unsafe items where students will be working.

4. **Plan visits to adopted trees in fall, winter, spring**
   Your class should visit the class' adopted trees (usually two trees) at least once during each season. During the visit, students should look for evidence of seasonal change. Let students do the telling; your questions will help them focus on different parts of the trees. Here's an overview of what to look for each season.

   In the fall, look for any seeds, cones, or fruits. Pull down a twig, and carefully remove a leaf at its base for students to examine. Mark the twig by trying a string around it so students can observe how it changes through the seasons.

   In the winter, find the twig that you marked with string in the fall, and let each student examine it. Have students search for the buds, leaf scars, and growth rings on their trees' twigs. Discuss any change to the trees that students notice.

   In the spring, find the twigs that you marked with string in the fall. Look for evidence of growth—new green twigs, leaves, buds, or flowers. If the tree is flowering, cut one flowering twig, and bring it into the classroom. Look for the mark on the trunk showing where you measured the tree in the fall. Use string to measure the same spot. Compare the measurements noted in the scrapbook.

   If you have school during the summer months, look for flowers and fruit.

5. **Plan assessment**

Here are specific objectives to observe in this part.

- *Trees are living, growing plants.*

- *Trees change through the seasons.*

Focus on a few students each session. Record the date and a + or − on the *Assessment Checklist*.

**What do fall trees look like?**

**Materials for Step 2**
- *String*
- *Camera*
- *Scissors*

*fall*
*food*
*fruit*
*season*
*seed*

# GUIDING *the Investigation*
## Part 3: *Fall: Visiting Adopted Trees*

1. ## Review the fall weather calendar
   Review the fall weather calendar and add today's weather conditions and temperature. Ask students to describe what **fall** weather is like where you live.

2. ## Go outdoors
   Take the class outdoors to visit their adopted trees. Look for any seeds, cones, fruits, and leaves.

   Choose a growing twig low enough for the class to observe. Ask them to tell you about the twig. Carefully remove one leaf where it joins the twig. Have each student examine the location where the leaf was attached. Ask,

   ➤ *Is it wet or dry?*

   ➤ *What is the shape of the mark where the leaf was?*

   Mark the twig with a string so it can be located again for further observations through the seasons. Take a photo of each tree.

3. ## Return to class
   When you return to the classroom, have students help you add the materials they have gathered to the tree scrapbook. Include photographs of each tree. This is a good time to review and add new vocabulary to the word wall.

4. ## Review vocabulary
   Reinforce key vocabulary used to describe trees parts in the fall. One way is with a cloze review. You say a sentence, leaving the last word off, and students answer chorally. Here's an example.

   ➤ *This is the season when trees lose their leaves; it is called _____ .*

   S: Fall (or autumn).

   ➤ *In the fall, many trees lose their _____ .*

   S: Leaves.

   ➤ *Fruits that come from trees are kinds of _____ .*

   S: Food.

   ➤ *Apples and oranges come from trees. They are examples of _____ .*

   S: Fruits.

5. **Focus question: What do fall trees look like?**

Write the focus question on the chart as you read it aloud.

➤ *What do fall trees look like?*

Tell students that you have a strip of paper with the focus question written on it for each of them. Have them glue the strips into their notebooks and use pictures and words to answer the focus question. When students finish, have them draw a circle so they can draw a close-up view of a leaf or twig. Visit each student to listen to and read his or her description.

**FOCUS CHART**

*What do fall trees look like?*

*Leaves change colors. Some trees lose their leaves.*

# WRAP-UP

6. **Share notebook entries**

Conclude the fall parts by having students share notebook entries. Ask students to open their science notebooks and read the focus question together.

➤ *What do fall trees look like?*

Ask students to pair up with a partner to

• share their answers to the focus question;

• explain their drawings.

## MATERIALS *for*

### Part 4: *Winter: Evergreen Hunt*

**For each student**

    1  *FOSS Science Resources: Trees and Weather*
- "Orange Trees"

**For the class**

- Evergreen leaves, needles, and scales (See Step 4 of Getting Ready.) ★

12–16 Zip bags, 1 L (or a laminator, optional) (See Step 4 of Getting Ready.) ★

    1  Big book, *FOSS Science Resources: Trees and Weather*

**For assessment**

- *Assessment Checklist*

★ Supplied by the teacher.

# GETTING READY for
## Part 4: Winter: Evergreen Hunt

1. **Schedule the investigation**

   Plan 5–10 minutes to introduce the investigation and 20 minutes for an outdoor evergreen hunt. Plan 15 minutes for the reading.

2. **Preview Part 4**

   Students hunt for evergreen trees that match samples of needles from schoolyard trees. The focus question is **What do winter trees look like?**

3. **Select your outdoor site**

   Survey the evergreen trees in your community that you would like students to visit. It is always a good idea to check the outdoor site on the morning of an outdoor activity. Check for any distracting items or unsafe items where students will be working.

4. **Make evergreen samples**

   Inventory the evergreen trees in your schoolyard. These may include conifers (pine, fir, spruce, hemlock, cedar, cypress, juniper, redwood) or broadleaf evergreens (magnolia, eucalyptus, palm).

   **Laminate and cut**

   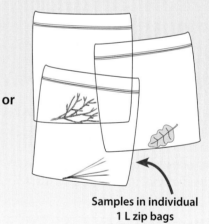

   **or**

   **Samples in individual 1 L zip bags**

   Select a total of 12–16 representative samples of leaves, needle clusters, or scaly twigs from your evergreen trees. You will need a sample for each pair of students. If you have only one kind of evergreen tree, select all the samples from that one kind of tree. If you have a wealth of evergreens, select samples from different kinds of trees. Laminate the samples to make reference cards (no paper is needed). Or put each of the samples into its own zip bag.

5. **Plan to read *Science Resources*: "Orange Trees"**

   Plan to read "Orange Trees" during a reading period near the end of this part.

6. **Plan assessment**

   Here is a specific objective for this part.

   - *Some trees lose their leaves in winter; others do not.*

   Focus on a few students in each session. Record the date and a + or – on the *Assessment Checklist*.

> **TEACHING NOTE**
>
> *The hunt should take place after the leaves have fallen but before snow prevents students from visiting trees up close. It also requires that your schoolyard have more than one evergreen species.*

## FOCUS QUESTION
**What do winter trees look like?**

## TEACHING NOTE

*This activity is best conducted after all the leaves have fallen from the deciduous trees. It can be effective, however, after deciduous trees have changed from green to their fall colors. Then, the emphasis should be on green/no green rather than leaves/no leaves.*

### Materials for Step 4
- *Evergreen samples*

## TEACHING NOTE

*The words* needle *and* scale *are introduced more fully to students in the context of the tree they visit during Step 6.*

## GUIDING *the Investigation*
### Part 4: *Winter: Evergreen Hunt*

1. **Discuss winter trees**
   Call students to the rug. Ask them to think about trees outdoors.

   *When we think about trees in the **winter**, we usually picture trees without leaves.*

   ➤ *Are all the trees leafless? Are any trees still green?*

   Tell students that they are going on a hunt to find trees that keep their leaves year-round. Explain that trees that have leaves on them through the winter are called **evergreen** trees. Discuss why this word seems appropriate.

2. **Go outdoors**
   Make sure students are dressed appropriately for a trip outdoors. Gather your evergreen reference samples and trek out to your home base in the usual orderly manner.

3. **Discuss** *evergreen*
   Ask students to look around for green trees. When green trees are pointed out, ask and explain,

   *Do you remember the name for trees that are green all the time? Trees that are green all year are called evergreen trees. Evergreen trees stay green because they have green leaves all year. Some trees change color and lose their leaves. But evergreen trees have green leaves all the time.*

4. **Introduce the evergreen hunt**
   Show students your leaf samples.

   *I have samples of leaves from some evergreen trees. Do you think any of the trees in our schoolyard have leaves like these?*

   Give each pair of students one of the evergreen samples and tell them,

   *This is a hunt. See if you and your partner can find a tree that has leaves, **needles**, or **scales** like the sample you have. When you find the tree, stand by it until I come to check.*

5. **Let the hunt begin**
   Send students off to find the trees that match their evergreen samples. As soon as a pair of students finds a match, visit them and confirm that the foliage is the same as their sample.

Needles

Scales

Ask the first team to follow you to the next tree to confirm that team's match. Continue visiting trees until you have gathered up everyone and you are together at the last evergreen tree.

6. **Discuss evergreen trees**

At the last tree, call for attention and discuss leaf type. Hold up the sample of leaves that match the evergreen tree you are under. Ask,

➤ *Is this an evergreen tree? How do you know?*

Describe the different kinds of foliage, focusing on their properties.

*Some evergreen trees have regular leaves that are wide and flat.*

*Some evergreen trees have leaves that are long, skinny, and pointed. These leaves are called needles. Pine, fir, and spruce trees have needles.*

*Some evergreen trees have tiny, pointy leaves stuck flat on the stems. These leaves are called scales.*

*Trees that have needles or scales are called conifers. Conifer trees also have cones, like pinecones.*

➤ *Does this evergreen tree have regular leaves, needles, or scales?*

➤ *How are these (needles, scales, or leaves) different from the leaves on our other evergreen trees?*

7. **Return to class**

As a group, visit each different kind of evergreen tree to look closely at the foliage and to compare it to other evergreens.

After completing your evergreen tour, organize students and return to the classroom in the usual orderly manner.

> **TEACHING NOTE**
>
> *Students might wonder if evergreens ever lose their needles. Instead of providing an answer, ask them to look for any evidence around the trees.*
>
> *Evergreens do lose their needles, leaves, and scales, just not all at the same time.*

# READING *in Science Resources*

8. **Read "Orange Trees"**

   This is the second of three articles about common trees through the seasons. Orange trees are evergreen and don't lose their leaves in the winter. In fact, the fruit on the navel orange tree turns orange with cooler temperatures in winter.

   Read the article aloud. Pause to discuss any tree structures that are mentioned in the reading.

9. **Discuss seasons and trees**

   After the reading, ask,

   ➤ *What season is it now where we live?* [Winter.]

   ➤ *We have been recording the weather on our calendar. What kind of weather have we had this winter?*

   ➤ *What does an orange tree look like in winter?* [Lots of green leaves with orange fruit on the tree.]

   ➤ *What season comes after winter?* [Spring.]

   ➤ *How do you think spring weather will be different from winter weather?*

   ➤ *What will happen to the orange tree in spring?* [Flowers will grow on the tree. Those flowers will become oranges later.]

**Orange Trees**

It is warm in winter where orange trees grow.
These oranges will soon be ready to eat.

40

# MATERIALS *for*
## Part 5: *Winter: Twigs*

### For each student at the center

1   Twig (See Step 3 of Getting Ready.) ★

### For the class

1   Book, *Our Very Own Tree*

2   Tree-trunk rounds

2   Loupes/magnifying lenses

5   Twig cross sections (See Step 4 of Getting Ready.) ★

1   Sharp knife ★

1   Pruning shears (optional) ★

1   Tree scrapbook

•   Water ★

•   White glue ★

•   Transparent tape ★

❏   1   Teacher master 28, *Center Instructions—Winter Twigs*

### For assessment

•   *Assessment Checklist*

★ Supplied by the teacher.          ❏ Use the duplication master to make copies.

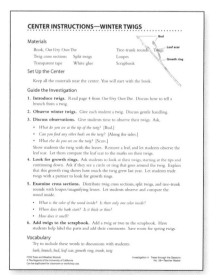

*No. 28—Teacher Master*

## GETTING READY *for*
### Part 5: *Winter: Twigs*

1. **Schedule the investigation**

   Work with groups of six to ten students. Plan about 20 minutes for each group at the center and 15 minutes for the reading.

2. **Preview Part 5**

   Students observe the inside of tree twigs and look for growth rings, buds, and leaf scars. The focus question is **What do winter trees look like?**

3. **Cut winter twigs**

   Check with your school grounds manager, a nursery, or friends to find out if they know of any trees being pruned.

   Cut 15 or more twigs from leafless deciduous trees. Cut twigs about 30 centimeters (12") long. Each twig should have an end bud and at least one growth ring. Look for twigs with easily identifiable leaf scars. Cut one twig that still has a few leaves attached.

   Set the twigs in water until needed. If other classes at your school are doing this investigation, share the twigs to avoid cutting too many from local trees.

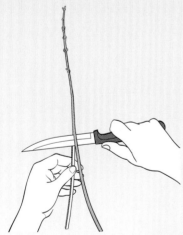

4. **Cut twig cross sections**

   Dissect five twigs. Use pruning shears or a sharp knife to cut across the butt ends of five twigs about 5 cm from the end. Save the cut pieces.

   Split the remaining parts of the five twigs lengthwise. Poke a knife through the center of the twig and wiggle it back and forth. Carefully move the knife along the length of the twig until the entire twig is split in half.

5. **Plan assessment**

   Here are specific objectives for this part.

   - *Bark, twigs, leaves, buds, flowers, fruits, and seeds are parts of trees.*

   - *Twigs have structures such as leaf scars and buds.*

   Focus on a few students each session. Record the date and a + or − on the *Assessment Checklist*.

# GUIDING *the Investigation*
## Part 5: *Winter: Twigs*

1. **Read *Our Very Own Tree***
   At the center, read page 4 of *Our Very Own Tree*: "We came up very close to the trunk." Help students think about the parts of the tree—from trunk to branches to twigs.

2. **Observe the winter twigs**
   Bring out the twigs. Discuss gentle handling of the twigs (they will be used again with each group). Give each student a twig, telling him or her to observe the twig closely.

3. **Discuss twig observations**
   After a few minutes of observation, ask,

   ➤ *What do you see at the tip of the twig?* [**Bud**.]

   ➤ *Can you find any other buds on the twig?* [Along the sides.]

   ➤ *What else do you see on the twig?* [**Scars**.]

   Show students the twig that still has leaves. Remove one of the leaves, and let each student take a close look at the location where the leaf stem was attached to the twig. Ask,

   ➤ *What shape is the mark where the leaf was attached?*

   Introduce the term *leaf scar* to describe the mark on the twig where a leaf was once attached.

   ➤ *Do you see leaf scars on your twig?*

4. **Observe twigs closely**
   Have students look closely at their twigs, starting at the tip and continuing down. Ask if they see a circle or ring that goes around the twig. Explain that this is the **growth ring** that shows how much the twig grew last year.

5. **Trade twigs**
   Have students trade twigs. Let them look for buds, leaf scars, and growth rings.

**Materials for Steps 2–3**
- *Twigs*
- *Twig with leaves*
- *Loupes/magnifying lenses*

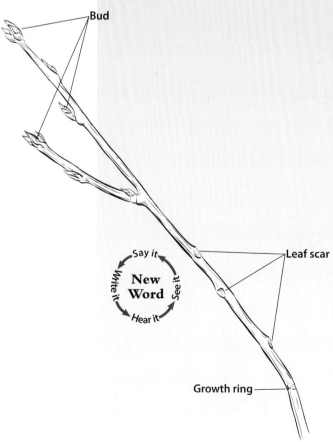

Bud

Leaf scar

New Word

Growth ring

**Materials for Steps 6–7**
- *Tree-trunk rounds*
- *Loupes/magnifying lenses*
- *Cut and split twigs*
- *Scrapbook*
- *White glue*
- *Tape*

## EL NOTE

*Draw a picture of the twig and label the parts on the word wall.*

6. **Examine twig cross sections**

   Show students the twigs that you have split, the small cross sections from the twigs, the tree-trunk rounds, and the loupes/magnifying lenses. Have students compare the wood inside the twigs and the rounds. Ask,

   ➤ *What color is the wood inside?*

   ➤ *Is there only one color inside?*

   ➤ *Where does the bark start? Is it thick or thin?*

   ➤ *How does it smell?*

7. **Add twigs to the scrapbook**

   Use a new page in the scrapbook to record your study of twigs. Glue in and tape a winter twig. Have some students help you label parts or dictate sentences to tell what they think is important to record in the scrapbook. Save room to add a spring twig later.

# MATERIALS *for*

## Part 6: *Winter: Visiting Adopted Trees*

### For the class

- 1   Weather calendar
- 1   Garden thermometer
- 1   Demonstration thermometer
- 1   Tree scrapbook
- &bull;   White glue ★
- &bull;   Transparent tape ★
- 1   Camera ★

### For assessment

- &bull;   *Assessment Checklist*

★ Supplied by the teacher.

## GETTING READY *for*
### Part 6: *Winter: Visiting Adopted Trees*

1. **Schedule the investigation**

   Plan 10 minutes for an introduction and to work with the weather calendar and scrapbook, 15 minutes outdoors for students to observe the adopted trees, and 10 minutes to draw and write in their notebooks in the classroom.

2. **Preview Part 6**

   Students revisit their adopted trees to observe any changes to the twigs, leaves, and areas around their trees. The focus question is **What do winter trees look like?**

3. **Check the outdoor site**

   It is always a good idea to check the outdoor site on the morning of an outdoor activity. Check for any distracting items or unsafe items where students will be working by their adopted trees.

4. **Plan assessment**

   Here are the specific objectives for this part.

   - *Trees change through the seasons.*

   - *Trees are living, growing plants.*

   Focus on a few students each session. Record the date and a + or − on the *Assessment Checklist*.

# GUIDING *the Investigation*
## Part 6: *Winter: Visiting Adopted Trees*

1. **Review the winter calendar and tree scrapbook**
   Review the winter weather calendar and add today's weather conditions and temperature. Ask students to describe what winter weather is like where you live.

   Bring out the scrapbook and review the pages of the trees that students adopted.

2. **Go outdoors**
   Take the class to visit their adopted trees. Take the scrapbook along so students can compare what the trees looked like in the fall to what they look like now. Discuss changes, and take a new picture of each adopted tree to paste into the scrapbook.

   Find the string you tied to a twig in the fall. Let each student examine the twig, looking for growth rings, leaf scars, and buds. Discuss any differences students notice since their fall visit.

3. **Revisit the class tree**
   If you planted a tree in the fall, visit it again to take pictures and make observations.

4. **Return to class**
   After returning to the classroom, have students help you add the materials they have gathered to the tree scrapbook. Include the photographs of each tree. This is a good time to review and add new vocabulary to the word wall.

5. **Review vocabulary**
   Reinforce key vocabulary used during Parts 4–6. One way to do this is to use a cloze review, where you say a sentence, leaving the last word off, and students answer chorally. Here's an example of cloze review.

   ➤ *The coldest season, when many trees are bare, is called _____ .*

   S: Winter.

   ➤ *Trees that keep their leaves through the winter are called _____ .*

   S: Evergreen.

   ➤ *Conifers are evergreen trees that have needles or scales. The long and thin ones are _____ .*

   S: Needles.

---

## FOCUS QUESTION
**What do winter trees look like?**

**Materials for Step 1**
- *Weather calendar*
- *Scrapbook*
- *Thermometer*

**Materials for Step 2**
- *Camera*

**Materials for Step 4**
- *White glue*
- *Tape*

bud
evergreen
growth ring
leaf scar
needle
scale
winter

➤ *The short and jointed ones are _____.*

S: Scales.

➤ *The mark where a leaf came off a twig is called a _____.*

S: Leaf scar.

➤ *The circle that goes around the twig to show how much it grew last year is called a _____.*

S: Growth ring.

➤ *The mark on the twig where a new leaf is forming is called a _____.*

S: Bud.

6. **Focus question: What do winter trees look like?**
Write the focus question on the chart as you read it aloud.

➤ *What do winter trees look like?*

Distribute focus-question strips for students to glue into their notebooks. Students should use pictures and words to answer the focus question.

## WRAP-UP

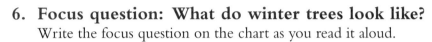

7. **Share notebook entries**
Conclude the winter parts by having students share notebook entries. Ask students to open their science notebooks and read the focus question together aloud.

➤ *What do winter trees look like?*

Ask students to pair up with a partner to

• share their answers to the focus question;

• explain their drawings.

---

**FOCUS CHART**

*What do winter trees look like?*

*Some trees are only branches. Some trees are evergreen.*

---

# MATERIALS *for*

## Part 7: *Spring: Forcing Twigs*

### For the class

- 10  Twigs ★
- 1  Tree scrapbook
- •  Transparent tape ★
- •  Soft-drink bottles, 2 L (See Step 4 of Getting Ready.) ★
- •  Gravel or pebbles ★
- 1  Scissors ★
- •  Water ★
- ❏  1  Teacher master 29, *Center Instructions—Forcing Twigs*

### For assessment

- •  *Assessment Checklist*

★ Supplied by the teacher.　　❏ Use the duplication master to make copies.

*No. 29—Teacher Master*

## GETTING READY *for*
### Part 7: *Spring: Forcing Twigs*

1. **Schedule the investigation**
   Plan 10–15 minutes for each group of six to ten students at the center.

2. **Preview Part 7**
   Students bring twigs into the warmth of the classroom to force them to bloom or put out leaves. The focus question is **What do spring trees look like?**

3. **Cut spring twigs**
   Cut two or three twigs from as many different kinds of trees as you can find *before* they start to flower or leaf out. Cherry, apple, willow, and plum are particularly good. Cut the twigs to include the end buds. Keep the twigs in water.

4. **Make twig vases**
   Make tall vases by cutting off the tapered top of 2 L soft-drink bottles, using scissors. Put about 5 cm (2") of gravel or pebbles in the bottles for stability. You'll need one vase for each type of twig.

5. **Plan assessment**
   Here are specific objectives for this part.

   - *Bark, twigs, leaves, buds, flowers, fruits, and seeds are parts of trees.*
   - *The buds on twigs grow into leaves or flowers.*

   Focus on a few students each session. Record the date and a + or − on the *Assessment Checklist*.

# GUIDING *the Investigation*

## Part 7: *Spring: Forcing Twigs*

1. **Observe the spring twigs**

   At the center, give each student a spring twig to observe. Bring out the scrapbook, and let students compare the **spring** twigs to the winter twigs. Ask,

   ➤ *How are these twigs different from those observed in the winter?*

   ➤ *Can you find new growth rings?*

   ➤ *Can you find the end bud?*

   ➤ *Can you find other buds?*

   ➤ *Are the buds the same size as they were in the winter? Are they bigger or smaller?*

2. **Sort twigs in vases**

   Have students sort twigs from one kind of tree and put them in a soft-drink-bottle vase. Ask students what they think might happen.

   The water and warmth of the room will stimulate some twigs to develop— a process called **forcing**. Buds may continue to swell, and the bud scales may fall off. The buds may open to reveal flowers or leaves. Discuss changes that students notice over time.

3. **Get ready for the next group**

   You might want to return the twigs to one of the vases before the next group arrives at the center so all students have the opportunity to sort the twigs into types. With the last group, tape one twig to the outside of each vase for later comparison.

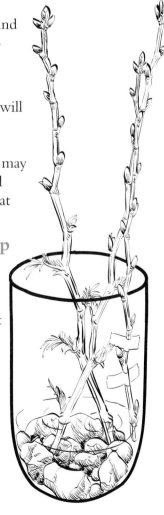

**FOCUS QUESTION**

*What do spring trees look like?*

**Materials for Step 1**
- *Twigs*
- *Soft-drink-bottle vases*
- *Scrapbook*
- *Water*
- *Tape*

**TEACHING NOTE**

*Over time, students will observe changes to the twigs in the water. Introduce the new vocabulary as they make the observations (*swollen *and* blossom*).*

Say it → See it → Hear it → Write it →
**New Word**

## MATERIALS *for*
### Part 8: *Spring: Bark Hunt*

For each student

    1   Bark photo (See Steps 3–4 of Getting Ready.) ★

For the class

    1   Tree scrapbook

    1   Camera ★

    •   Crayons (optional) (See Step 6 of Getting Ready.) ★

    •   White paper (optional) ★

   32   Zip bags, 1 L (or a laminator, optional)

For assessment

    •   *Assessment Checklist*

★ Supplied by the teacher.

# GETTING READY *for*
## Part 8: *Spring: Bark Hunt*

1. **Schedule the investigation**

   Plan on 5–10 minutes for an introduction and 20–25 minutes for outdoor tree-bark observations. Plan on 10 minutes for tree scrapbook work back in the classroom.

2. **Preview Part 8**

   Students observe and compare bark on a variety of trees as they search for matches to photos of the bark on schoolyard trees. The focus question is **What do spring trees look like?**

3. **Select your outdoor site**

   Walk around your schoolyard and determine the several trees that you will use for the bark matching activity. Look for trees with obvious differences (color, texture, bark pattern).

4. **Photograph bark**

   Take close-up photographs of the bark of the selected schoolyard trees. Each student will need one photo.

   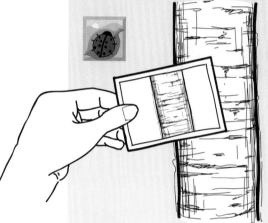

   Print an equal number of copies of each bark pattern, making sure that the total number of photos equals or exceeds the number of students. (Several students will have copies of the same photo.) To protect the photos, laminate them or put them in zip bags.

5. **Check the site**

   It is always a good idea to check the outdoor site on the morning of an outdoor activity. Check for any distracting items or unsafe items where students will be working.

6. **Plan for bark rubbings (optional)**

   If you plan to have students make bark rubbings, you will need white paper and crayons. Peel the paper from the crayons before heading outdoors. If you didn't make rubbings in Investigation 1, practice making a rubbing before you have students try.

7. **Plan assessment**

   Here are specific objectives for this part.

   - *Bark, twigs, leaves, buds, flowers, fruits, and seeds are parts of trees.*

   - *Trees can be identified by the pattern of the bark.*

   Focus on a few students each session. Record the date and a + or − on the *Assessment Checklist*.

**Materials for Step 2**
- *Bark photos*

**TEACHING NOTE**

*If students cannot act autonomously, move from tree to tree as a group, giving each group an opportunity to compare and declare a match or no match.*

**Materials for Step 4**
- *White paper*
- *Peeled crayons*

# GUIDING *the Investigation*
## Part 8: *Spring: Bark Hunt*

1. **Introduce the bark hunt**
   Call students to the rug. Ask,

   ➤ *What is on the outside of a tree?* [Bark.]

   ➤ *Is all bark the same?*

   ➤ *What is the texture of the bark? How does the bark feel?* [Rough.]

   ➤ *Have you ever smelled the bark of a tree?*

   Tell students that they will go on a bark hunt. Show them the photos you have taken of bark from trees around the schoolyard.

2. **Distribute bark photos**
   Distribute a bark photo to each student. Ask students to find others who have a photo exactly like theirs. Once reorganized into groups, students should sit down together. Tell students,

   *We are going outdoors now. You have a picture of the trunk of one of the trees in our schoolyard. Do you think you can figure out which tree it is? Let's go see if we can figure out which tree you have.*

3. **Go outdoors**
   Organize the class into their leaving-the-classroom configuration and travel to your outdoor home base in the usual orderly manner. At home base, describe the challenge once again.

   - *Stay together as a group.*

   - *Walk to a tree with your bark picture.*

   - *Compare the bark on the tree with the picture.*

   - *If the picture matches the bark, stand by the tree.*

   - *If the bark does not match, walk together to a new tree.*

   When a group declares a match, join them to determine if they have made a correct comparison. Have the group join you as you all walk to the next group. Continue until you have gathered up everyone and you are together at the last tree.

4. **Make bark rubbings (optional)**
   Review (or demonstrate) how to use paper and crayons to make rubbings of the bark of a tree.

   Some students will have trouble holding the paper still as they make a bark rubbing on a tree trunk. Show students how to work with a partner, taking turns holding the paper tightly against the tree trunk while the other rubs with the crayon.

Have students compare the bark rubbings to the pictures of the bark and discuss the similarities and differences.

Students may want to do a rubbing from a leaf of the same tree to help identify the tree, as bark rubbings can look very similar. A leaf rubbing is best done in the classroom where smooth, flat surfaces are available.

5. **Return to class**

After completing the rubbings, file back to your classroom in the usual orderly manner.

Call students to the rug. Add photographs and rubbings of the spring tree bark to the class scrapbook. Ask students to contribute words to describe the appearance of the bark.

## MATERIALS *for*
### Part 9: *Spring: Visiting Adopted Trees*

For each student

   1   *FOSS Science Resources: Trees and Weather*

     •  "Maple Trees"

For the class

   1   Weather calendar

   1   Garden thermometer

   1   Demonstration thermometer

   •  String

   •  Circumference strings (from Investigation 1, Part 5)

   2   Tree-trunk rounds

   1   Tree scrapbook

   1   Camera ★

   •  White glue ★

   •  Transparent tape ★

   1   Scissors ★

   1   Big book, *FOSS Science Resources: Trees and Weather*

For assessment

   •  *Assessment Checklist*

★ Supplied by the teacher.

# GETTING READY *for*

## Part 9: *Spring: Visiting Adopted Trees*

1. **Schedule the investigation**

   Plan 5–10 minutes for an introduction, 15 minutes outdoors for students to observe their adopted trees, and 10 minutes indoors or outdoors to draw and write in their notebooks. In addition, plan 10 minutes for the reading.

2. **Preview Part 9**

   Students revisit their trees. They look for evidence of new growth in the leaves and flowers. The focus question is **What do spring trees look like?**

3. **Check the outdoor site**

   It is always a good idea to check the outdoor site on the morning of an outdoor activity. Check for any distracting items or unsafe items where students will be working by their adopted trees.

4. **Plan to read** *Science Resources*: **"Maple Trees"**

   Plan to read "Maple Trees" during a reading period near the end of this part.

5. **Plan assessment**

   Here are specific objectives for this part.

   - *Trees change through the seasons.*

   - *Seasons change in a predictable annual pattern: fall, winter, spring, and summer.*

   - *Trees are living, growing plants.*

   Focus on a few students each session. Record the date and a + or − on the *Assessment Checklist*.

**Materials for Step 1**
- *Weather calendar*

**Materials for Step 2**
- *String*
- *Camera*
- *Scissors*

**Materials for Steps 3–4**
- *Circumference strings*
- *Tree-trunk rounds*
- *White glue*
- *Tape*
- *Scrapbook*

blossom
flower
forcing
spring
swollen

# GUIDING *the Investigation*
## Part 9: *Spring: Visiting Adopted Trees*

1. **Review the spring calendar**

   Review the spring weather calendar and add today's weather conditions and temperature. Ask students to describe what spring weather is like where you live.

2. **Go outdoors**

   With the class, return to the adopted trees to check the twigs that you marked in the fall. End buds may be swelling, or leaves and blossoms may be appearing. New twig growth may be noticeable by its lighter green color.

   Find the circumference mark on the trunk and measure it with string again. Take pictures of the adopted trees to add to the scrapbook. If the tree is flowering, cut one flowering twig and bring it into the classroom.

3. **Return to class**

   Back in the classroom, compare fall and spring trunk-measurement strings. Bring out tree-trunk rounds once more. Point out the area between the bark and the inner wood. Tell students that this is where the tree grows. The rings are layers of wood that the tree has added each year.

4. **Add to scrapbook**

   Have students help you add the materials they have gathered to the tree scrapbook. Include the photographs of each tree. This is a good time to review and add new vocabulary to the word wall.

5. **Review vocabulary**

   Reinforce key vocabulary used during Parts 7–9. Here's an example of cloze review.

   ➤ *New leaves and flowers grow on trees in the season called _____.*
   S: Spring.

   ➤ *New leaves and flowers come out of the _____.*
   S: Buds.

   ➤ *Some trees have only leaves, but some have leaves and _____.*
   S: Flowers.

6. **Focus question: What do spring trees look like?**
Write the focus question on the chart as you read it aloud.

➤ *What do spring trees look like?*

Distribute a strip of paper with the focus question for students to glue into their notebooks. They should use pictures and words to answer the question.

**FOCUS CHART**

*What do spring trees look like?*

*Trees have flowers and new leaves growing on their twigs and branches.*

# WRAP-UP

7. **Share notebook entries**
Conclude the spring parts by having students share notebook entries. Ask students to open their science notebooks and read the focus question together.

➤ *What do spring trees look like?*

Ask students to pair up with a partner to

- share their answers to the focus question;
- explain their drawings.

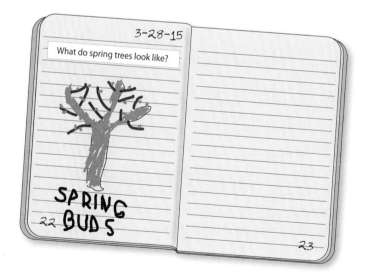

3-28-15

What do spring trees look like?

SPRING
22  BUDS

23

## READING *in Science Resources*

8. **Read "Maple Trees"**

   This is the third and final article about common trees through the seasons. Read it aloud while students follow along in their own books. Ask them to find the maple tree in summer, fall, and winter on page 45.

   On page 46 there is a close-up photo of maple seeds. The article asks,

   ➤ *What will happen when they are planted and grow?*

   Have students respond to this question about the maple seeds. Confirm that seeds will grow if they are planted and that the maple seeds will grow up to be maple trees.

9. **Discuss seasons and trees**

   After the reading, ask,

   ➤ *What season is it now where we live?* [Spring.]

   ➤ *We have been recording the weather on our calendar. What has the weather been like this spring?*

   ➤ *How did our adopted trees change in the spring?*

   ➤ *How did our adopted trees change in the fall?*

   ➤ *How did our adopted trees change in the winter?*

   ➤ *What do you think our adopted trees will look like in summer?*

   ➤ *What do you predict our adopted trees will look like next fall, winter, and spring?*

   Review the word *pattern* and the types of patterns students have seen in this module. They may have experienced repeating patterns in mathematics with shapes or colors. Tell them that seasons have a pattern, too. Ask them to describe the pattern of the seasons from fall to summer.

   Ask students if they can think of another repeating pattern that they experience every day of the year. If they have trouble, ask them if day and night is a repeating pattern of events.

**Maple Trees**

In spring, maple trees have new green leaves.
Can you name the **season** for each maple tree?

44

# INTERDISCIPLINARY EXTENSIONS

## Art Extension

- **Make a tree bulletin board**
  Dedicate part of a bulletin board to a large paper tree. Begin with a large trunk and branches. Cover the trunk with bark rubbings. Have students make leaves. Change the tree as students notice things happening to trees throughout the seasons.

## Science Extensions

- **Watch for seed showers**
  Trees will bud, flower, develop fruit, and scatter their seeds. With some trees, this can happen with a burst, filling the air with floating seeds. When the trees begin to let go of their seeds, it's time for a seed-collecting walk.

- **Watch for life in trees**
  In the spring, have students watch for signs of life in their trees. Are ants crawling up the trunks? Where are they going? Where are they coming from? If birds are building nests, students could draw pictures of the nests being built, the birds guarding the nest, and the young birds that appear.

- **Compare cones**
  Collect a variety of cones for students to explore. Old, brittle, half-open cones can be cracked open to reveal the seeds with their delicate winged cases. Soak open cones overnight, and they'll close up to protect any hidden seeds.